Molecular Biology:
An Introduction to
Chemical Genetics

Concepts of Modern Biology Series

William D. McElroy and Carl P. Swanson, *Editors*

CONCEPTS OF MODERN BIOLOGY SERIES

William D. McElroy and Carl P. Swanson, *Editors*

Molecular Biology: An Introduction to Chemical Genetics

J. M. Barry

University Lecturer
and Fellow of St. John's College
University of Oxford, England

and

E. M. Barry

Prentice-Hall, Inc., Englewood Cliffs, New Jersey

Library of Congress Cataloging in Publication Data

BARRY, JOHN MICHAEL.
  Molecular Biology.

  (Concepts of modern biology series)
  Bibliography: p.
  1. Molecular genetics. I. Barry, E. M., joint
author. II. Title.
QH431.B302 1973   574.8'732   72-8232
ISBN 0-13-599522-1
ISBN 0-13-599514-0 (pbk)

© 1973 by PRENTICE-HALL, INC.
Englewood Cliffs, New Jersey

10 9 8 7 6 5 4 3 2 1

Printed in the United States of America

PRENTICE-HALL INTERNATIONAL, INC., *London*
PRENTICE-HALL OF AUSTRALIA, PTY. LTD., *Sydney*
PRENTICE-HALL OF CANADA, LTD., *Toronto*
PRENTICE-HALL OF INDIA PRIVATE LTD., *New Delhi*
PRENTICE-HALL OF JAPAN, INC., *Tokyo*

To M. J. McC.

Contents

Foreword

Concepts of Modern Biology Series   The main body of biological literature consists of the research paper, the review article or book, the textbook, and the reference book, all of which are too limited in scope by circumstances other than those dictated by the subject matter or the author. Unlike their usual predecessors, the books in this series, CONCEPTS OF MODERN BIOLOGY, are exceptional in their obvious freedom from such artificial limitations as are often imposed by course demands and subject restrictions.

Today the gulf of ignorance is widening, not because of a diminished capacity for learning, but because of the quantity of information being unearthed, most of which comes in small, analytical bits, undigested and unrelated. The role of the synthesizer, therefore, increases in importance, for it is he who must take giant steps, and carry us along with him; he must go beyond his individual observations and conclusions, to assess his work and that of others in a broader context and with fresh insights. Hopefully, the CONCEPTS OF MODERN BIOLOGY SERIES provides the opportunity for decreasing the gulf of ignorance by increasing the quantity of information and the quality of presentation. We are convinced that such volumes occupy an important place in the education of the practicing and prospective teacher and investigator.

WILLIAM D. MCELROY and CARL P. SWANSON

Preface

The purpose of this book is to introduce, in a simple way, discoveries in chemical genetics—discoveries that show precisely how inherited differences between living organisms are founded on differences in the structure of molecules. Since 1964, when the first edition of this book appeared, a new generation of students has arisen for whom molecules are the natural approach to biology, and who show no surprise that genetics should be explicable in chemical terms. To provide these students with a worthwhile introduction, we have rewritten the first edition, not only describing new work, but also raising slightly the level of treatment throughout.

J. M. Barry
E. M. Barry

Chapter 1

Genes: Structures Within Cells that Control Heredity

I. Gregor Mendel

Genetics is the study of heredity in living organisms, that is, their tendency to beget like organisms. This tendency is shown most clearly by single-celled creatures. By division an *Escherichia coli* bacterium, for example, gives two progeny that have appearance and properties identical to the parent. How can this be explained? At first sight the answer might seem obvious. We know that the characteristics of the *E. coli* bacterium, like those of all other living things, depend solely on the molecules of which it is composed. Hence the explanation would seem to be that during cell division the component molecules and structures formed from them are distributed evenly between the daughter cells.

This explanation, however, is inadequate because the daughter cells are half the parental size and must double in size before they in turn divide. It is this doubling of cell size that is the basic problem of heredity, both in single-celled and multicellular organisms. Since cells are made of molecules, the increase in size must normally involve a doubling in the number of each kind of molecule in the cell. It is possible to conceive of various ways in which this process might come about, but all ways involve some form of self-copying

1

by certain molecules of cells. There might conceivably be only one kind of molecule in the cell that could form copies of itself, and this might direct the formation, or entry into the cell, of all other molecules. Or there might be a number of molecules that could form self-copies. Or one kind of molecule might catalyze the copying of another, which in turn might catalyze the copying of the first.

Such an understanding of heredity in chemical terms is the task of chemical genetics and, as this book will show, it has now largely been attained. The first clues to a correct understanding came not from studies of cell molecules or the division of single-celled organisms but from Gregor Mendel's experiments in the nineteenth century on heredity in pea plants. Heredity in such higher organisms is less obviously related to the replication of cell molecules than it is in single-celled organisms, but it is, once again, founded on it. For example, flower color in peas, whose inheritance Mendel studied, results from cells of the flower forming molecules of a certain pigment. This pigment was not present in the pollen or egg cell that united to give the fertilized egg from which the flowers were derived. But the pigment must have been formed by some molecule, or molecules, that were present in the pollen or egg cell (or both) and that were exactly copied after each of the cell divisions that led to the cells of the flower.

Mendel was, in effect, the founder of chemical genetics, and the outline of his life is well known. He was born in 1822, the son of a farmer, and at twenty-one became an Augustinian monk. He lived for most of his life at the Altbrünn Monastery in Brünn, which was then in the Austrian Empire. It was in the monastery garden that he carried out his experiments on inheritance in peas, and although the experiments were published in 1866 in the Brünn scientific journal, their significance was not grasped until after his death.

This familiar outline, however, tends to romanticize Mendel's life, and it does not provide a clear explanation of his brilliant scientific ability. In fact, he was not a monk in the sense of being a member of a large enclosed community. His monastery contained only about thirteen priests, who were chosen for their intellectual ability. They were free to travel and receive guests, and the monastery was an important intellectual center with an excellent cuisine. Figure 1 shows Mendel, at the time of his experiments, with his fellow monks. The abbot, in the front row, was an oriental scholar of repute, and also included are a famous authority on Goethe and a noted composer and expert on church music. Mendel entered the monastery with a good education and was then sent to the University of Vienna for two years to study science. After his return, he taught biology for fourteen years at a secular high school in Brünn, with an excellent staff and over a thousand boys. Reminiscences of his pupils show that he was a friendly, humorous man and a good teacher. It was during this period that Mendel performed

Figure I. Mendel and his fellow monks. He is standing one from the extreme
right. (Courtesy of Mrs. Anne Iltis.)

his experiments. He was later elected abbot of his monastery. It is clearly no
more remarkable that Mendel should have performed successful experiments
in his monastery garden than that at the same time Charles Darwin should
have done so in the garden of his home near London.

2. Mendel's Theory of Inheritance

The phenomenon of heredity is an obvious fact of life, but before Mendel
announced the results of his experiments no satisfactory explanation of it
had been given. A popular theory was that both a father and a mother
contributed to their offspring some kind of fluid that contained an essence
of their characteristics. The blend of the two fluids determined the characteristics of the offspring, which tended to be intermediate between those of
the two parents. But a number of facts did not fit with this theory, and it was
not generally accepted by scientists. For example, if a black sheep is crossed
with a white ram, their offspring are not gray but pure white, and if these
breed among themselves, some of their offspring will be white and some
black. In the early nineteenth century numerous experiments had been
performed on plant hybridization—the crossing of plants that differ from
one another. But, as Mendel pointed out at the beginning of his paper, there

had been "formulated no generally applicable law governing the formation and development of hybrids." He added, "Those who survey the work done in this department will arrive at the conviction that among all the numerous experiments made, not one has been carried out to such an extent and in such a way as to make it possible to determine the number of different forms under which the offspring of hybrids appear, or to arrange these forms with certainty according to their separate generations, or to definitely ascertain their statistical relations." This he intended to remedy. His paper describes a precise and elegant series of experiments on hybridization in peas and advances a simple theory to explain the results. He implies that his theory could explain the facts of heredity in general.

Mendel's theory of heredity will be explained first, and the experiments from which he derived it will be described later. The theory is really made up of two parts: a theory of precisely how one plant differs from the next in its inherited character, and a theory of how this character originates in each plant. The first part suggests that the general character that any pea plant inherits can be subdivided into a large but limited number of characteristics, which can be called *unit characters*. Flower color is an example of a unit character. Plants that differ in their inherited character do so in possessing one or more of these characters in sharply different forms. One plant, for example, can have red flowers and another white. This revolutionary theory suggested how a precise analysis could be made of differences in character that had previously been considered only in a general and descriptive way.

We now know that Mendel's theory that an individual is a mosaic of distinct characters is correct. It extends to all living organisms and is one of the basic principles of biology. It implies that the possible number of genetically different individuals in a species, though large, is limited. For example, genetically different human beings result from the alternative forms of the thousands of unit characters being combined together in different ways. If all these ways were exhausted, any further individuals would be identical twins of those that had come before. The theory, in effect, claims that inherited character is built up from a number of fundamental particles. If heredity is to be explained in chemical terms, this is a necessary consequence of the particulate theory of matter.

The second part of Mendel's theory explains how the particular form in which each character will appear is determined. He suggested, in effect although not in so many words, that every egg and pollen cell of the pea plant carries one determinant for each character. The fertilized egg therefore carries two determinants for each character, and the form in which the character will appear is decided by their interaction. For example, every egg and pollen cell carries a determinant for flower color. If both cells that combine to form the fertilized egg carry the determinant for red, the flowers will be red, and if for white, the flowers will be white. If one cell carries the determinant for red and the other carries that for white, the flowers, as will be seen later, will be red. Mendel did not consider the nature of the determinants. After the rediscovery of his work, it was suggested by some biologists that they had no reality, but were merely convenient figments for explaining the facts of heredity. The suggestion parallels another made in the nineteenth century that atoms and molecules have no physical reality. It is now clear that Mendel's determinants, which have been named genes, have as real an existence as do the atoms and molecules of which they are composed, and that they control the appearance of characters in all living organisms.

3. Mendel's Evidence in Support of his Theory

Chemists of the eighteenth century discovered that matter is composed of particles largely because they made precise measurements of the quantities of compounds that react together, rather than merely describing the appearances of chemical reactions. Similarly, Mendel discovered the particulate nature of inheritance largely because he made precise counts of the number of plants that showed different characters, rather than merely describing them as his predecessors had done. Mendel chose pea plants for his experiments for two reasons: because they possess sharp differences in numerous points of character, and because their flowers are normally self-fertilized and receive no pollen from without unless this is introduced experimentally. He selected strains of plants that differed sharply in one or more of seven characteristics and that maintained these differences through generations of self-fertilization. The seven characteristics are examples of what we shall call the unit characters of his theory. Their alternative forms were round or wrinkled seed, yellow or green seed, red or white flowers, smooth or wrinkled pods, green or yellow pods, flowers distributed along the stem or bunched at the end, and stems of 6 to 7 feet or of $\frac{3}{4}$ to $1\frac{1}{2}$ feet.

He first made crosses between these plants to produce hybrid seed. He found that when the parent plants differed in a certain character, one of the alternative forms always appeared in the hybrid to the exclusion of the other. For example, when the pollen from round-seeded plants was used to fertilize the flowers of wrinkled-seeded plants, or vice versa, the seeds that formed from these flowers were always round. Or, when red-flowered plants were crossed with white-flowered ones and the seeds that formed were grown into new plants, these plants all had red flowers. Mendel called round seed and red flowers the *dominant* forms of these two unit characters. The dominant forms of the other characters were yellow seed, smooth pods, green pods, flowers along the stem, and stems of over 6 feet.

Mendel then grew large numbers of each of these seven kinds of hybrid seed into plants that he allowed to self-fertilize, and he then collected the seeds that formed. Some of these seeds, or the plants that were grown from them, showed the forms of the characters that had disappeared in their parents. For example, Mendel grew 253 hybrid round-seeded plants and allowed them to self-fertilize. In the pods, which formed on the plants, were usually both round and wrinkled seeds. The wrinkled form of the character had obviously been latent in the hybrid plant, and he called it the *recessive* form of the character.

Findings of this kind—the disappearance of a characteristic in a cross and its reappearance in subsequent generations—had been observed by plant breeders before Mendel. Darwin, for example, described experiments on crossing two forms of *Antirrhinum* with normal and abnormal flowers. The first generation all had normal flowers, but 88 of their progeny had normal, 37 abnormal, and 2 intermediate flowers. Darwin was not led by these figures to Mendel's comprehensive experiments with some 30,000 pea plants and merely concluded that the tendency to abnormal flowers "appeared to gain strength by the intermission of a generation." Mendel's important discovery resulted from his counting the numbers of different forms of each character. This revealed remarkably simple ratios between them. For example, from the hybrid round-seeded plants that had been allowed to self-fertilize he collected 7,324 seeds; 5,474 were round and 1,850 were wrinkled, a ratio of 2.96 to 1. From large numbers of each of the six other kinds of hybrid plants he obtained, after self-fertilization, the dominant and recessive forms in the ratios of 3.01, 3.15, 2.95, 2.82, and 3.14 to 1. All these ratios, as Mendel perceived, are nearly 3 to 1, and they most closely approached this ratio when large numbers of plants were counted.

This ratio provides beautiful evidence of the truth of Mendel's theories that were discussed in the last section. He implied that every egg and pollen cell contains one determinant for each character and that the fertilized egg, therefore, contains two. In a hybrid fertilized egg the determinants for at least one character will differ. When the resulting hybrid plant forms egg cells or pollen cells, one half will contain the determinant for one form of this character, and the other half will contain that for the other. For example, half of both the pollen and egg cells from the hybrid round-seeded plants will contain the determinant for round seed and half that for wrinkled. When these cells fertilize one another, "it remains," in Mendel's words, "purely a matter of chance which of the two sorts of pollen will become united with each separate egg cell." It is like the simultaneous tossing of two coins: there is one chance that two heads will fall together, to two chances that a head will fall with a tail, to one chance that two tails with fall together. Hence, after a large number of self-fertilizations, one quarter of the eggs will contain two determinants for round seed, one quarter will contain two for wrinkled, and one half will contain one for each. Only the eggs that contain two determinants for wrinkled seed will in fact become wrinkled seed, and hence there will be a ratio of 3 to 1 between round and wrinkled seed. The round seeds should be of two kinds. When grown into plants, one third should, on self-fertilization, give only round seed, while two thirds should give round and wrinkled in the ratio of 3 to 1. Mendel tested these round seeds and proved that this was in fact so.

Mendel also performed experiments to test another point: in effect, whether the determinants for different characters are linked together in any way. For example, he had two strains of plants that had always bred true and that differed in two unit characters. One gave seeds that were both round and yellow, and the other gave seeds that were both wrinkled and green. Were the determinants for round and yellow seed and those for wrinkled and green seed linked together, or could they separate freely? To discover the answer, he crossed plants of the two strains together, and, as expected, the resulting seeds were all round and yellow. He grew these seeds into plants and allowed them to self-fertilize. He had proved that half the pollen and egg cells from these plants contain the round determinant and half the wrinkled, and, also, that half contain the yellow determinant and half the green. Did those cells that contained the round determinant contain only the yellow, because these two determinants were linked together, or did half contain the yellow and half the green? If the two pairs of determinants were in fact linked, this self-fertilization would give only round, yellow seeds and wrinkled, green seeds, in the ratio of 3 to 1. Mendel in fact found all four combinations, often in the same pod. Out of 556 seeds, 315 were round and yellow, 101 wrinkled and yellow, 108 round and green, and 32 wrinkled and green. The ratio between these is 9.8 to 3.2 to 3.4 to 1. If the determinants for the two characters separated quite independently, the ratio should be 9 to 3 to 3 to 1 (Figure 2). Mendel concluded from this that they do move independently and that deviations between the experimental and theoretical ratios were due to chance.

He confirmed this conclusion in another way. He again grew the hybrid round and yellow seeds into plants, but instead of allowing them to self-fertilize, he made a *backcross* with the strain with wrinkled and green seed— the strain, that is, whose pollen and egg cells contain only the recessive wrinkled and green determinants. If the round and wrinkled determinants are in fact unlinked to the yellow and green determinants, the hybrid plants should produce eggs and pollen grains of four kinds in equal numbers: those with one round and one yellow determinant, those with one wrinkled and one yellow, those with one round and one green, and those with one wrinkled and one green. A backcross with the doubly recessive plants should then produce the four kinds of seed in equal numbers.

When Mendel pollinated the hybrid plants with pollen from the doubly recessive strain, he obtained 31 round and yellow seeds, 27 wrinkled and yellow, 26 round and green, and 26 wrinkled and green. When doubly reces-

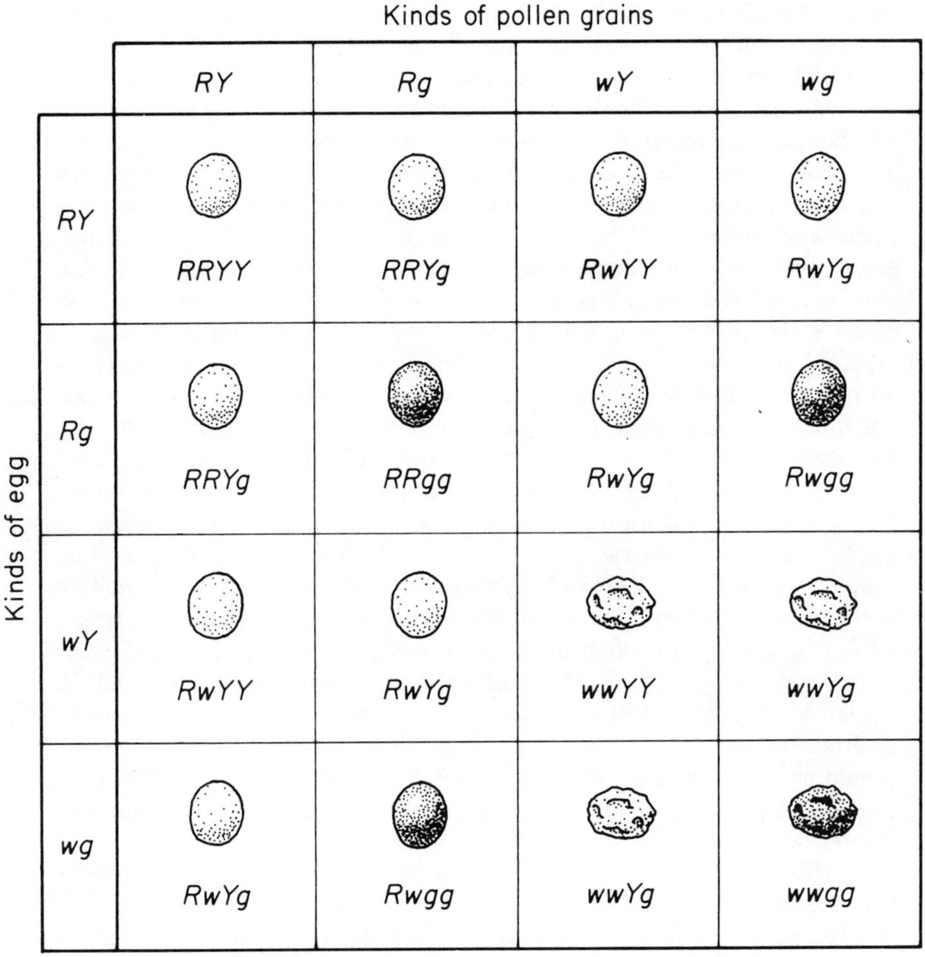

Kinds of pollen grains

	RY	Rg	wY	wg
RY	RRYY	RRYg	RwYY	RwYg
Rg	RRYg	RRgg	RwYg	Rwgg
wY	RwYY	RwYg	wwYY	wwYg
wg	RwYg	Rwgg	wwYg	wwgg

Kinds of egg

Figure 2. Origin of 9:3:3:1 ratio. The hybrid plants that were self-fertilized produced eggs and pollen grains with one determinant for each character in the four combinations shown. These four kinds of germ cells were formed in equal numbers. Random pollination produced fertilized eggs with two determinants for each character in the 16 combinations shown. Sixteen kinds of seed were thus produced in equal numbers, but they appeared to be of only four kinds, as illustrated.

sive plants were pollinated by the hybrid plants, he got 24 round and yellow seeds, 22 wrinkled and yellow, 25 round and green, and 26 wrinkled and green. It is clear that these numbers were close enough to being equal to confirm his conclusion that these determinants move independently. It will be seen later that this backcross (or *testcross*, as it is now usually called) with a doubly recessive individual is the normal method used to reveal the genetic constitution of the germ cells of a hybrid organism. Its value lies in the fact that each type of germ cell from the hybrid individual produces a visibly different offspring.

Mendel also performed other comprehensive experiments that strongly support his theories. When reading his paper it is painful to think of the unrewarded emotional energy that must have gone into his experiments on peas, which took him seven years to complete. It is not surprising that in his later years he seems to have had a mild persecution complex.

4. Mendel's Paired Determinants Lie on Paired Chromosomes

If Mendel's determinants really exist, where do they lie in the germ cells, and do they also occur in all other cells of an organism? Mendel did not pursue these questions, and they were not considered until his work was rediscovered eighteen years after his death. The reason the significance of his work was not grasped immediately is difficult to determine. His paper was read by a number of leading biologists (though apparently not by Darwin) and it was quoted in books on hybridization. Although he was politely commended for his thorough techniques, nobody grasped what he was driving at. Part of the reason is probably that his experiments and theory, though extremely lucid, are somewhat shrouded in modest and abstract language. Also, he tended to express himself in mathematics, which most biologists are not quick to grasp. For example, he summarizes his experiments that show that determinants for different characters are not linked by saying, "the offspring of the hybrids in which several essentially different characters are combined represent the terms of a series of combinations, in which the developmental series for each pair of differentiating characters are associated." This is preceded by an algebraic expression occupying five lines. Above all, it is probable that his thinking was unfashionable and that, in their initial enthusiasm for evolution by natural selection, biologists were unsympathetic to a theory that suggested that new individuals could arise solely by the reassortment of the fixed characters of their parents. However convincing Mendel's numerical ratios may have appeared at first sight, most biologists probably suspected them of hidden flaws because they could not be integrated into contemporary theories.

Between 1866, when Mendel's results were published, and 1900, when their significance was recognized, important advances occurred in biological knowledge and thought. During the 1880s, details of cell division were clarified. The cell nucleus, like the cell itself, was seen to divide, and this division was named *mitosis*. The deeply staining material of the nucleus was named *chromatin*. At the start of mitosis, when the nuclear membrane disappears, the chromatin was seen to condense into separate threads, which were named *chromosomes*. Each chromosome appeared to divide longitudinally at mitosis to give two identical chromosomes that separated and became part of the nuclei of the two daughter cells. Moreover, it was demonstrated that at fertilization there is a fusion not only of egg and sperm cells but also of their nuclei.

Thus the central feature of cell division appeared to be the even distribution of chromosomes to the daughter cells. This fact suggested to many biologists that the chromosomes contain a material that determines the heredity of the cell, that is, its characteristic structure and function. August Weismann named this material *germ plasm* and developed his "theory of the continuity of the germ plasm." He suggested that the germ plasm of a fertilized egg remains unchanged in structure along the line of cell divisions leading to the egg or sperm of an adult organism. Weismann proposed that the egg and sperm make equal contributions to the germ plasm of the fertilized egg. He suggested that chromosomes are the carriers of the germ plasm and predicted that the number of chromosomes must be halved at a nuclear division during the formation of each egg and sperm in order to prevent an increase in chromosome number from generation to generation. This "reduction division" was soon observed under the microscope.

By 1900, then, the known facts of biology made it far easier to perceive where Mendel's observations and theory could fit in the puzzle of heredity. Mendel's theory requires that his determinants be present in pairs in the fertilized egg and in many other cells. But when the germ cells are formed, these determinants must separate, only one of each pair passing into a germ cell.

In 1900 three biologists, Hugo De Vries, Carl Correns and Erich von Tschermak, independently made discoveries similar to Mendel's, and after completing their experiments discovered a reference to Mendel's work. Soon after this W. S. Sutton, a graduate student at Columbia University, published a paper on the chromosomes of a species of grasshopper in which he suggested that "the association of paternal and maternal chromosomes in pairs and their subsequent separation during the reducing division . . . may constitute the physical basis of the Mendelian law of heredity." He showed that during the *meiotic* cell divisions during the formation of germ cells when the number of chromosomes per cell is halved, a chromosome originally derived from the organism's father pairs with one identical in shape and size that was originally derived from the mother. He gave good evidence that when these paired *homologous* chromosomes separate into the two new cells it is a matter of chance which member of each pair goes into which cell. Hence, if in peas the determinants for round and wrinkled seed were carried on one pair of homologous chromosomes and those for yellow and green seed on another, this random separation of paired chromosomes would allow the independent assortment of these characters found by Mendel. It thus became evident that the chromosomes contain not an ill-defined "germ plasm" but Mendel's discrete determinants, and the road to molecular genetics lay open.

A flood of experiments followed, and in a few years character differences had been found in over 100 plants and 100 animals that were inherited like the character differences of Mendel's peas. Some of the most important work was done by T. H. Morgan and his colleagues at Columbia University. Morgan chose the fruit or vinegar fly, *Drosophila melanogaster*, for his genetic experiments because it breeds rapidly and is easy to look after. He bred large numbers of flies in milk bottles and in only a few months could perform experiments similar to those that had taken Mendel several years. Morgan soon showed that Mendel's discoveries with pea plants applied to *Drosophila*. Thus, normal and vestigial wings appeared to be alternative forms of a single unit character, the appearance of which resulted from the interaction of a pair of Mendel's determinants (which were now named *genes*). When normal flies were crossed with those with vestigial wings, their progeny were all normal. But when these were crossed with one another, vestigial wings reappeared in their progeny, and the ratio between the dominant and recessive forms was 3 to 1.

The alternative forms of a gene determining a single unit character became known as *alleles*. An individual bearing identical alleles for a certain character became known as *homozygous*, and one bearing different alleles as *heterozygous*. It became clear from work on *Drosophila* and other organisms that one allele is not always dominant over the other: the character can appear in a form that is intermediate between the two extremes. Thus when black domestic fowls were crossed with white ones, their progeny (known as the F_1 generation) were not all black or all white but all gray. When the gray fowls were crossed with one another their progeny (the F_2 generation) consisted of approximately one black homozygote to one white homozygote to two gray heterozygotes.

Morgan and others also found, as Mendel had, that the genes for certain different characters were not linked together. Thus when flies with smooth eyes and gray bodies were crossed with others with rough eyes and black bodies, their progeny had smooth eyes and gray bodies. But when these were crossed with one another to give the F_2 generation, the four possible combinations of characters were found in the ratio of 9 to 3 to 3 to 1 (see p. 8).

However, as Sutton pointed out, the number of unit characters in an organism must greatly exceed the number of chromosomes. Hence if genes do in fact lie in chromosomes, many genes should be linked together and not show the independent assortment found by Mendel. Such *linkage* was in fact discovered. In *Drosophila*, for example, when flies with normal eyes and normal wings were crossed with others with purple eyes and vestigial wings, their progeny all had normal eyes and wings. When these were crossed with one another, only the two original combinations appeared in *most* of the progeny and in a ratio of 3 to 1. Clearly these genes controlling eye color and wing size are joined together and only occasionally separate into different germ cells.

However, there were at first difficulties in accepting linkage as evidence that genes lie on chromosomes: in addition to the linkage being broken with varying frequency, in some organisms the number of characters that segregated independently exceeded the number of chromosomes. For example, the pea has 7 pairs of chromosomes, yet 11 independently segregating characters were found, including the 7 studied by Mendel. This dilemma was only solved over a number of years by a thorough study, largely by T. H. Morgan and C. B. Bridges, of linked characters. As described in the next section, it became clear that the genes of an organism fall into a number of pairs of linked groups equal to the number of pairs of chromosomes. Deviations from complete linkage are explained by exchange of material between paired chromosomes, and certain pairs of genes, although borne on the same pair of chromosomes, are so far apart that they are always separated by this exchange during the formation of germ cells and hence behave as if not linked.

5. Arrangement of Paired Genes on Paired Chromosomes: Mapping Techniques in Higher Organisms

In addition to proving conclusively that genes are part of chromosomes, the study of linkage provided another important clue to the chemical structure of genes: it showed that they are arranged one after another as if along some kind of thread within the chromosome.

In the last section two linked genes controlling eye color and wing size in *Drosophila* were mentioned. This linkage was sometimes found to be broken: when flies homozygous for Normal eyes and Normal wings were crossed with others homozygous for purple eyes and vestigial wings, a few flies in the F_2 generation had Normal eyes and vestigial wings and a few others had purple eyes and Normal wings. (We have written the dominant forms with a capital first letter.) In other words, the ratio between the four possible combinations of these characters was neither 3:1:0:0 as it would be for complete linkage nor 9:3:3:1 as it would be if there were no linkage, but it lay between the two. By crossing the abnormal F_2 flies with one another, it became clear that the alleles for Normal eyes and vestigial wings were now linked together as were those for purple eyes and Normal wings. Hence in a few of the germ cells of the F_1 generation the original linkage had been broken and the genes had undergone *recombination*.

The most probable explanation of recombination was that homologous chromosomes exchange segments sometime before the reduction division, and conclusions about the arrangement of genes on chromosomes were deduced on this assumption. Later, in 1931, the assumption was proved correct by H. B. Creighton and B. McClintock at Cornell University. They studied recombination in a strain of maize between two genes that were known to be carried on a particular pair of homologous chromosomes. These chromosomes had come to differ from one another in microscopic structure: one had a knob at one end and a fragment from another chromosome attached to the other end. Whenever maize plants arose whose characteristics showed that recombination between the particular genes on this chromosome had occurred, microscopic examination showed that the knob and attached fragment were now on different members of the homologous pair. In the same year similar experiments were reported with *Drosophila*.

The exchange of segments between homologous chromosomes, or *crossing over*, must occur sometime before they separate at the reduction division in the formation of germ cells. A complication arises here that makes it necessary to outline the behavior of chromosomes in the two meiotic divisions. At the first meiotic division, homologous chromosomes become aligned side by side, before separating into two new cells. But it can be seen that each homologous chromosome is itself made up of two identical chromosomes (or *chromatids*) lying side by side. Hence each chromosome has duplicated sometime before pairing. In the second meiotic division the chromatids of each chromosome move into separate cells, which thus have half the normal number of chromosomes. In males all four of the cells produced by the two meiotic divisions usually form active sperms, but in females only one forms an egg. Therefore, exchange of segments between homologous chromosomes occurs sometime before they separate, and the chromosomes also duplicate before they separate. Does exchange of segments occur before or after duplication?

It would seem most fitting if exchange of segments occurred during the pairing of homologous chromosomes peculiar to meiosis, that is, after duplication. Moreover, at this stage, adhesions can usually be seen between pairs of chromatids (Figure 3). These *chiasmata* (singular, *chiasma*) look very much like the result of exchange of segments between one chromatid and another, but for many years it was extraordinarily difficult to verify this conclusively. Hence the possibility remained that exchange of segments

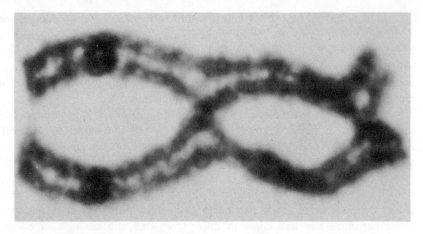

Figure 3. Pair of homologous chromosomes of the salamander at meiosis. Each chromosome consists of two chromatids. The two black circles at the left between pairs of chromatids are centromeres. In the center and again at the extreme right are chiasmata. (Courtesy of Dr. James Kezer.)

occurred at some earlier stage of the cell cycle. However, combined studies of inheritance and meiosis in *Drosophila*, and in fungi, have shown conclusively that exchange occurs at the *four-strand stage* between one chromatid and another. The frequency of chiasma formation also correlates with the frequency of recombination between genes, and it seems certain that chiasmata are a visible consequence of exchange between chromatids.

In spite of the lack of conclusive microscopic evidence for crossing over, study of recombination in the early part of the twentieth century gave important information about the arrangement of genes on chromosomes. This information was founded on precise measurements of *recombination frequency*: the proportion of sperms or eggs formed by an organism in which recombination has occurred between two particular genes. The usual method of measuring recombination frequency is by means of a testcross, first used by Mendel (p. 9). Bridges, in fact, made this testcross in *Drosophila* for the two linked characters that we have been considering and published his results. He first crossed females homozygous for Normal eyes and Normal wings with males homozygous for purple eyes and vestigial wings. The progeny all displayed the dominant Normal characters, since in their cells each recessive allele was paired with its dominant allele. Bridges first took males from this F_1 generation and testcrossed them with homozygous recessive females. All the eggs of these females contained one allele for purple eyes and one for vestigial wings. Half of the sperm of the males contained one allele for Normal eyes and half contained one for purple eyes, and half contained one allele for Normal wings and half contained one for vestigial wings. The testcross was designed to find in what proportion of these sperms the original linkage had been broken so that they contained both the allele for Normal eyes and that for vestigial wings or the allele for purple eyes and that for Normal wings. The result of the testcross was 519 flies with Normal eyes and Normal wings (produced from sperms with these alleles) and 552 flies with purple eyes and vestigial wings (produced from sperms with these alleles). None of the flies had purple eyes and Normal wings or Normal eyes and vestigial wings. Hence the linkage between these genes for eye color and wing shape had never been broken during the formation of the sperms of the F_1 males.

However, when Bridges made the same testcross with the F_1 females and homozygous recessive males, the result was different. The number of flies with Normal eyes and Normal wings was 1,339 and the number with purple eyes and vestigial wings was 1,195. But there were also 151 flies with Normal eyes and vestigial wings (produced from eggs with these alleles) and 154 with purple eyes and Normal wings (produced from eggs with these alleles) (Figure 4). Hence at sometime during the formation of 2,839 $(1,339+1,195+151+154)$ eggs of the F_1 females, recombination between these genes for eye color and wing shape had occurred 305 $(151+154)$ times. Hence the recombination frequency was 10.7% $[(305/2,839)\times 100]$. In the previous cross the recombination frequency had been zero—an illustration of the fact that although recombination occurs in the formation of eggs in *Drosophila*, it never occurs in the formation of sperms. In most animals and plants this is not so: recombination between any pair of genes occurs with roughly equal frequency in the formation of sperms and eggs. Recombination frequency is only slightly affected by variations in environment such as temperature.

The fact that recombination never occurs during the formation of *Drosophila* sperms makes it easy to assign *Drosophila* genes to linkage groups by a testcross. The number in *Drosophila melanogaster*, which has four pairs of chromosomes, was found to be four. Moreover, the relative number of genes in each group was roughly the same as the relative sizes of the four chromosomes. Two other species of *Drosophila* are known with five and six pairs of chromosomes and they have five and six linkage groups. In all organisms studied to the present date, the number of linkage groups equals the number of pairs of chromosomes—strong support for the belief that genes lie on chromosomes.

Morgan found that recombination frequencies between different pairs of genes in a linkage group differed in an interesting way: any three genes between which the recombination frequencies were small could be arranged in an order such that the sum of the recombination frequencies between genes *A* and *B* and genes *B* and *C* equaled that between *A* and *C*. For example,

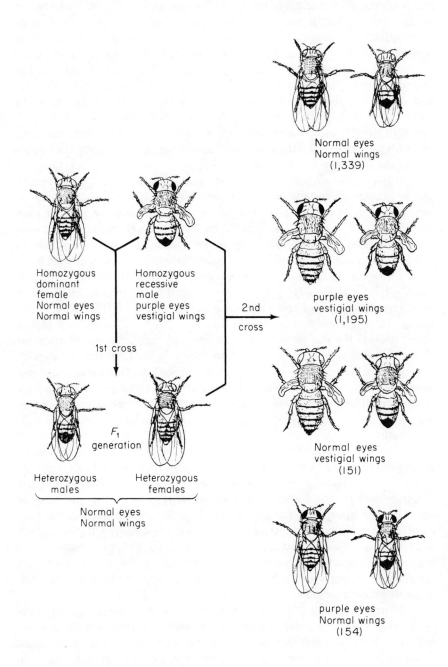

Figure 4. Testcross in *Drosophila*.

one experiment showed the recombination frequency between the genes *crossveinless* and *cut* (both wing characters) to be 7.9% and that between *cut* and *vermillion eye* to be 15.2%; the frequency between crossveinless and vermillion was 22.7%, or almost exactly the sum of the two.

The only simple explanation of such evidence was that genes are arranged linearly along the chromosome and that recombination between linked genes results from exchange of segments between paired homologous chromosomes or chromatids. The reason for this may be illustrated as follows. Suppose that we had two chains each composed of 26 "plug-in" necklace beads, which represent a pair of homologous chromatids that will separate at the reduction division. Suppose that we mark the beads in one chain with the capital letters A to Z to represent 26 dominant genes and in the other with the lowercase letters a to z to represent recessive alleles. Then we lay the two chains straight out side by side, with the letters in the two chains correctly aligned. Then we break each chain at the same point and recombine the sections from different chains. Suppose that the breakage had been between the letters L and M on one chain and l and m on the other. Then one of the new chains will have A to L and m to z. The other will have a to l and M to Z.

Suppose that one such crossover occurred between each pair of homologous chromatids before the formation of each germ cell and that crossing over occurred at random along the length of the chromatids. Then the linkage between two genes at the extreme ends of the chromatid (A and Z and a and z, in our example) would be broken during the formation of every germ cell; that is, the recombination frequency would be 100%. Linkage between a gene situated at the end of the chromatid and one at the middle would be broken during formation of half of the germ cells (recombination frequency 50%), and so on. In our example of the necklace there are 25 links between the 26 beads, and hence for every crossover there is one chance in 25 that the link between two adjacent beads will be broken. Hence the recombination frequency between, for example, A and B or between B and C will be 4% and that between A and C will be $4\% + 4\% = 8\%$.

If this kind of explanation for recombination is correct, it should be possible to deduce the relative positions of genes on a chromosome. Such a chromosome map for *Drosophila* is shown in Figure 5. *Map distances* are shown beside each gene and are calculated by adding the recombination frequencies between every known pair of genes that precede it on the map. A glance at the map shows that our simple illustration of crossover between necklace beads is inadequate. In that illustration the map distance for a gene at the end of a chromosome would be 100, whereas two in this map are in fact over 100. Moreover, when the recombination frequency between two genes at either end of a chromosome map is measured experimentally, it never exceeds 50, and hence is less than the sum of the recombination frequencies between intermediate pairs of genes.

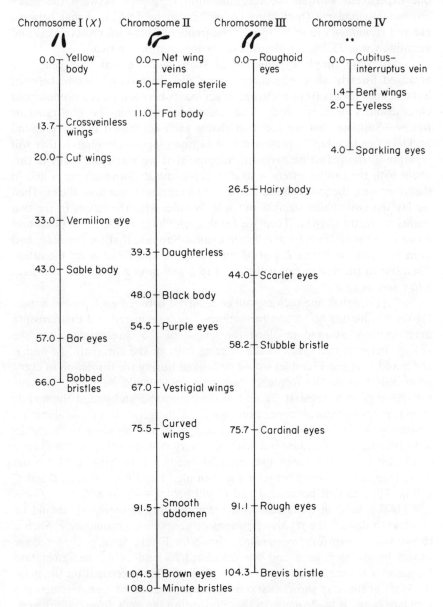

Figure 5. Map of some genes on chromosomes of *Drosophila melanogaster*. The appearance of each pair of chromosomes is shown at the top. Chromosome I is the X or sex chromosome. In males one of the two X chromosomes is replaced by a Y chromosome of different shape.

These discrepancies can be explained by multiple crossovers. If recombination between three linked genes ABC and abc is studied, a very few progeny can be found that have aBc or AbC linked on a single chromosome. This can only result from two crossovers, with the first breakage between A and B and a and b and the second between B and C and b and c. (In fact, such recombinants show at once that gene B lies between A and C, and this is the principal technique used to assign genes to their relative positions on a chromosome map such as Figure 5).

If the number of crossovers that occurred between chromatids each time two homologous chromosomes paired was fairly large and variable, there would be two important consequences. First, the recombination frequency between genes at either end of the chromosome would be 50%. This can again be illustrated by the necklaces: all odd numbers of crossovers between the two necklaces will separate beads A and Z, and all even numbers of crossovers will reunite them. Hence on the average, separation of A and Z will occur during half of the pairings of homologous chromatids, to give a recombination frequency of 50%. (Multiple crossovers in fact involve all four chromatids, but the explanation remains basically the same.)

Second, the sum of the recombination frequencies between successive pairs of genes would exceed 50%. Suppose, for example, that there are an average of five crossovers between our two necklaces of 25 links each time they pair. Then the linkage between any two successive beads will be broken and recombined, on the average, once in 25/5 = 5 pairings. The recombination frequency between successive beads is thus 20%, and the sum of all 25 recombination frequencies is 500%, even though the experimentally determined frequency between the terminal beads is only 50%. We can also understand in another way why the recombination frequency between distant genes is not the sum of the intermediate frequencies. Consider two successive pairs of beads AB and BC. In 20% of all pairings, crossover will occur between A and B, and in 20% it will occur between B and C. Hence in 20% of 20% (or 4%) of the pairings it will occur between A and B and between B and C simultaneously, and A and C will not be separated. Hence the measured recombination frequency between A and C will be 20%+20%−4% = 36%. Experimental evidence supports the belief that recombination between the genes of higher organisms can be explained by chromosomal behavior of essentially this type.

We have illustrated recombination by necklaces in which the beads represent genes. During the early work on recombination this analogy was thought to be accurate. Each gene was envisaged as a unique combination of molecules that in some way controlled a unit character; a slight structural change in these molecules could produce an allele. Genes were thought to be strung together by a genetically inert material that was broken during recombination. We shall see later that the analogy to beads is not, in fact, so precise: chromosome breakage during recombination can occur within genes as well as between them, and the linear arrangement of genes on a chromosome is paralleled by a linear arrangement of the structural material of the gene itself.

6. Arrangement of Genes on Unpaired Chromosomes: Mapping Techniques in Bacteria and Viruses

The rapid advances in genetics of recent years have been founded on the discovery that genes linked into chromosomes also control the individuality of bacteria and viruses, which produce vast numbers of progeny under simple conditions of culture. The genes of bacteria and viruses are linked into a single unpaired chromosome, and the form of each unit character results from the action of only one gene. (During rapid division more than one chromosome can accumulate in bacteria, but these are all derived from the same parent chromosome and have identical structures.) To conclude this chapter, techniques will be outlined that can reveal the arrangement of genes on the chromosomes of bacteria and viruses.

To describe the first technique for mapping genes on bacterial chromosomes, it is necessary to go back to the surprising discovery of sexual conjugation in bacteria, made in 1946 by J. Lederberg and E. L. Tatum at Yale University. It had been known for many years that bacteria multiply by cell division, and until 1946 it was believed that bacteria did not mate or come into any significant physical contact with one another. Therefore it was thought that the genes of any bacterium were derived from only one parent. Lederberg and Tatum showed that this is not so, and that occasionally two bacteria will mate with the formation of a bacterium whose single chromosome has a combination of the genes of both parents. They made this discovery essentially as follows.

They had two strains of *E. coli* that differed in a number of unit characters. The first was able to make its own supplies of the vitamin biotin and of the amino acids phenylalanine and cysteine but was unable to make the vitamin thiamine and the amino acids leucine and threonine, which had to be provided in the culture medium. The second could not make biotin, phenylalanine, and cysteine but could make thiamine, leucine, and threonine. Either strain when incubated alone, with the necessary nutrients, multiplied to give many more bacteria all of which had the same characters as the parent strain, except for a few cells whose genes for one of the characters had changed by mutation. But because mutation occurs only rarely and randomly, the chance of it occurring simultaneously in three genes in the same cell is extremely remote. Hence, on separate culture of the strains, bacteria did not normally appear that could grow in a medium without any vitamins or amino acids. When, however, the two strains were grown together, such bacteria did occasionally appear. Lederberg and Tatum concluded that genes were passing from some cells of one strain to some of the other and proved that this occurred by the cells coming into contact, rather than by transfer of the genes through the medium.

The details of this bacterial conjugation have been elucidated largely by F. Jacob and his colleagues at the Pasteur Institute in Paris. They proved that *E. coli* bacteria are of two types: donors (or males) and recipients (or females). Donor cells, in addition to their chromosome, have several copies of a separate particle named F, the fertility or sex factor. Donor cells are therefore designated F^+ and recipients F^-. Each F^+ cell has a surface component that makes it able to adhere to an F^- cell. A narrow tube then forms between the two cells. When F^+ and F^- strains are grown in the same culture, a high proportion of the cells mate in this way, and a sex factor passes into most of the mating F^- cells. But into about 1 in 10 million of the F^- cells genes from the F^+ cell pass (with or without the sex factor) and exchange with the corresponding genes of the F^- chromosome by recombination. This transfer is a consequence of the sex factor becoming attached to the F^+ chromosome. High-frequency-recombinant (*Hfr*) strains can be isolated in which the sex factor is more or less permanently attached to this chromosome. When these are incubated with F^- cells, about 1 in 10,000 of the F^- cells undergoes recombination with F^+ genes.

Jacob was able to make use of this mating between *Hfr* and F^- cells to reveal the arrangement of genes on the *E. coli* chromosome. He proved that, like the chromosomes of higher organisms, it is long and narrow, with the genes linearly arranged, and that during mating it passes from the *Hfr* donor into the F^- recipient, like a rope passing through a narrow hole. This proof depended on his finding that conjugating bacteria, which are normally attached to one another for about 30 minutes, could be broken apart by stirring the culture in a Waring blender—the machine often used as a household mixer.

In one experiment he allowed two strains of bacteria to mate that differed in six unit characters and hence in six genes. If he left them to mate for 30 minutes, recipient cells could be isolated that by recombination had gained all six donor genes. But if they were broken apart, after 25 minutes only five of the genes could be found in recipient cells. One which conferred the ability to ferment the sugar galactose had not passed in, as shown by the fact that no cells were formed that could grow in a medium with galactose as the only source of carbon. If separated after 18 minutes, both this gene and another controlling lactose fermentation were excluded; if after 11 minutes, these two and another conferring immunity to a virus; if after 10 minutes, these three and another conferring sensitivity to a drug; if after 9 minutes, these four and another conferring the ability to make the amino acid leucine; and if after 7 minutes, all these five genes plus the sixth, which conferred the ability to make the amino acid threonine, did not enter the recipient.

In all experiments with these two strains, different genes passed into the recipient in the same order in time. It was clear that the genes are linked together in this order into a chromosome resembling that of higher organisms. The sex factor was always the last to pass into the F^- cell, showing that it was attached at the end of the incoming chromosome. However, when a number of *Hfr* strains were studied, a surprising variation in the sequence of genes in the chromosome from one strain to the next was discovered. Suppose that the order in which the genes entered the recipient cell from one *Hfr* strain was *abcd...xyz* followed by the sex factor. From another *Hfr* strain

Figure 6. Map of some genes on *E. coli* chromosome. The difference between any two of the numbers gives the time in minutes taken by the segment between them to pass into a female cell during conjugation.

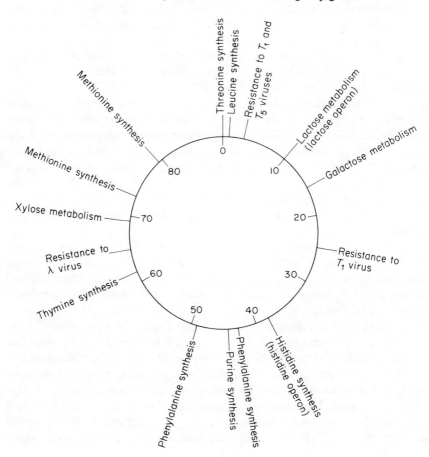

it might be *bcd . . . xyza* followed by the sex factor, and from another *efgh . . . xyzabcd* followed by the sex factor. It must be concluded that the genes in *E. coli*, unlike those in higher organisms, are arranged in a circle that can be opened at various points by the sex factor (Figure 6). It will be seen in a later chapter that the electron microscope shows that the *E. coli* chromosome is in fact circular.

The *interrupted mating technique* described above is one way in which the chromosome of *E. coli* may be mapped, but for genes that are closely linked, other techniques are needed. One of these is *recombination mapping*. When a portion of the chromosome of an *Hfr* cell enters an F^- cell, its genes are only expressed in the progeny if they are incorporated into the F^- chromosome by replacement of the F^- genes. Let us represent the F^- genes by the alphabet of capital letters (i.e., *ABC . . . XYZ* but with *Z* united to *A* to form a circle). Let us represent the incoming *Hfr* genes by lowercase letters. Suppose that a segment of the *Hfr* chromosome represented by *abcdefg* enters an F^- cell on conjugation. Recombination occurs much as in higher organisms: a segment of the F^- chromosome (say, *CDE*) is replaced by the homologous segment of the *Hfr* chromosome (i.e., *cde*). (The process differs from that in higher organisms in that two breaks must be made in the F^- chromosome rather than one—i.e., between *B* and *C* and between *E* and *F*.) On studying the characteristics of recombinant cells, we shall find that they more frequently contain *a* plus *b*, or *b* plus *c*, than they do *a* plus *c*. Hence their relative order is *abc*.

There are two other important techniques for the mapping of bacterial chromosomes: *transduction* and *transformation*. In transduction the donor bacterial cells are infected with a particular bacterial virus (or *bacteriophage*). Some of the progeny phages that are released when the bacterial cells burst have fragments of the bacterial chromosome attached. The phages are then used to infect the recipient bacterial strain. About 1 in 1 million phages introduce a fragment of the donor bacterial chromosome into a recipient cell in such a way that it is incorporated into the recipient chromosome by recombination, without the recipient being killed. As is usual in recombination, the closer two donor genes lie together on the donor chromosome, the more frequently they are incorporated into a single recipient chromosome; as a result the relative position of the genes on the donor chromosome can be determined. The transducing phage carries only about one hundredth of the donor chromosome (or about 50 genes) into the recipient cell. Hence transduction is useful only for mapping genes that lie near one another.

In transformation the chromosomal material is actually extracted by chemical techniques from the donor bacteria and added to a culture of the recipient bacteria. Some chromosomal fragments enter recipients and replace homologous portions of the chromosome by recombination. The extracted chromosomal fragments are normally somewhat smaller than those transferred in transduction, and hence transformation can again be used only for mapping genes that lie close together. The discovery of transformation is described later (p. 55) since analysis of the transforming material provided evidence for the chemical structure of genes.

The chromosomes of bacteriophages are mapped by infecting bacteria with two strains of phages that differ in their inherited characteristics. When the two kinds of phage multiply within a single bacterium, segments of the chromosome of one phage may be replaced by the homologous segment from another. As usual in recombination studies, the relative frequencies with which two or more genes are transferred from one strain to another can indicate their relative positions. A study of recombination in bacteriophage also provided important clues to the chemical structure of genes (see p. 64).

Chapter 2

The Molecular Structure
of Cells

1. Cell Structure and Vitalism

At the beginning of the last chapter it was promised that a clue to the fundamental problem of cellular heredity—the replication of the molecules of a cell after division—would be provided by Mendel's discoveries. But at first sight, the finding that inherited characteristics can be subdivided into unit characters, each controlled by one or two genes, appears to have little immediate relation to cell molecules. In particular, the various unit characters, such as plant height, seed shape, and flower color in peas, appear so unrelated that a common chemical basis is difficult to conceive. Although the rediscovery of Mendel's work revealed an ordered basis for heredity, how it could be explained in material terms remained mysterious. For example, in 1902 the famous geneticist W. Bateson wrote, in an enthusiastic account of Mendel's discoveries, "Let us recognize from the outset that as to the essential nature of these phenomena we still know absolutely nothing. We have no glimmering of an idea as to what constitutes the essential process by which the likeness of the parent is transmitted to the offspring.... We do not know what is the essential agent in the transmission of parental characters, not even whether it is a material agent or not. Not only is our ignorance complete, but no one has the remotest idea how to set to work on that part of the problem."

Gradually, however, the experiments of Mendel and his successors were seen to give a clue as to how the replication of molecules after cell division might occur. Living organisms are composed of molecules. Therefore, certain differences in the formation of cell molecules must underlie the different forms that each unit character can manifest. These differences, in accord with Mendel's experiments, result from the action of different genes. Genes, therefore, control the formation of at least some cell molecules. Hence the simplest explanation of cellular heredity would be that genes, directly or indirectly, control the formation, or entry into the cell, of all other molecules and that the molecules of which they are themselves composed are in some way capable of forming self-copies. This explanation has been largely verified by the discoveries of molecular biology, which will be described in the remaining chapters. We shall first consider the chemical structure of genes, which is a necessary prelude to explaining heredity in terms of chemistry. We shall then consider how genes form self-copies, and finally how genes control the formation of specific RNA and protein molecules. These molecules are the common basis of the diverse Mendelian unit characters, and through them, genes control the formation of other molecules of the cell.

This understanding of cell function is founded on studies of cell structure that have progressed from two directions. Over the last century chemists have gradually revealed the structure of larger and larger biological molecules, culminating in recent years in demonstrations of the bonding of every atom in certain protein and nucleic acid molecules. Parallel with these chemical studies of molecules of gradually increasing size has run the study by microscopists of cell structures of gradually diminishing size. The microscopes available around the middle of the nineteenth century revealed that cells of higher organisms are built on a common plan, with a spherical nucleus in the center surrounded by viscous cytoplasm. As the years passed, light microscopes of increasing magnification were made, and techniques for revealing cell structures by staining were refined. As a result the cytoplasm was seen to contain numerous particles. But the smallest cell particle discerned with the light microscope was enormous compared with the largest cell molecule whose structure had been discovered by the chemist. Between the two lay the mysterious viscous fluid of the cytoplasm in which other particles could be dimly discerned.

Moreover, it could be proved that the size of the waves of visible light is too great for light microscopes ever to reveal the fine structure of the cytoplasm. Having struck this evidently impenetrable barrier, some biologists came to view the cytoplasm almost mystically. To it were attributed such qualities as "irritability" and "self-perpetuation." The names *cytoplasm* and *protoplasm* did not help, since they suggested entities that were mysteriously more than the atoms and molecules of which they were composed. These vague vitalist emotions tended to inhibit clear thinking about the mechanism of living cells—although, as we shall see, some vitalist theories were clearly developed.

However, in recent years the studies of chemists and microscopists have met: protein and nucleic acid molecules whose structures are understood by the chemist can now be seen in the electron microscope, which has about 1,000 times the resolving power of the most powerful light microscope, and the smallest particles that can now be seen in the cytoplasm are composed of only a few protein and nucleic acid molecules. With this junction of chemists and microscopists a shaft of light has, in effect, pierced the cell. The mist that shrouded the cytoplasm and that evoked vitalist beliefs has been dispersed, and there remain no regions of the cell whose molecular structure cannot be envisaged.

Although many vitalist beliefs have been vague and irrational, by no means have all been. Many clear-minded scientists of the past, influenced by the complexity of living creatures, developed precise vitalist theories of a kind that could conceivably be put to experimental test. We shall now make a brief digression to explain these theories, since they define important possibilities that should be borne in mind when attempting to explain biology in molecular terms.

To explain the theories of vitalists, it is first necessary to state the theories with which they do not agree: those of mechanists. A mechanist believes that a living organism results from its component atoms and molecules being situated in the correct relative positions and that life is solely the interaction of these atoms and molecules according to the same laws of chemistry and physics as are displayed in the inanimate world.

Those theories of vitalists that could conceivably be tested by experiment are of two distinct kinds. Those of the first kind can be illustrated by the ideas of Justus von Liebig, a great German biochemist of the nineteenth century. In his textbook *Animal Chemistry*, he appears to agree with the mechanists that a living organism merely results from its component atoms and molecules being correctly situated relative to one another and that a living organism could, at least in theory, be made in the laboratory. But he believed that when the atoms and molecules of a living organism are in their correct relative positions, a new force of nature manifests itself. This vital force is the principal cause of the molecular reactions and movements at the basis of life. This force, he believed, is inherent in all atoms and molecules but only manifests itself in the complex structure of the cell, and research in biochemistry would reveal the laws of its action, as experiments in physics had revealed the laws of gravitation and electrostatic attraction.

The possibility of vital forces of this type was still entertained in 1945 by the famous physicist Erwin Schrödinger in his book *What Is Life?*, which made a great impact on scientists, and persuaded many biologists and physicists to study the genetics of bacteria and their viruses in the hope of discovering new laws of nature. Schrödinger compares the biologist to an engineer who is familiar with heat engines but encounters an electric motor for the first time. Although it is made of familiar materials, he suspects that laws that he does not understand are involved in its function. He nevertheless believes that these are understandable ordered laws. "He will not suspect that an electric motor is driven by a ghost because it is set spinning by the turn of a switch, without boiler and steam."

In contrast, vitalists of the second kind do believe in a ghostly force and that the basis of life lies outside the normal realm of science. Their theories are varied, but the most clearly defined have been those of Hans Driesch, who died in 1942. He started his career as an experimental biologist and studied the development of fertilized eggs into adult organisms and made important contributions to the subject, but he became convinced that this development would never be explained solely by the interaction of forces of nature of the kind normally investigated by scientists. As a result, he retired from experimental science early in this century and became a professor of philosophy at the University of Heidelberg, where he developed his theories of vitalism.

Driesch believed that living organisms differ sharply from nonliving in the possession of a vital force and that this force is unlike those familiar to scientists. It is a purposive directive force of a kind suggested by Aristotle. Its purpose is that a living organism should grow to maturity and reproduce. It acts by directing the component parts of the organism into paths they would not take under the sole influence of the normal forces of nature, so causing the organism to develop and function correctly. Driesch claimed that the existence of this force was suggested by numerous characteristics of living organisms, such as the resistance of animals to infection, and was proved beyond all reasonable doubt by certain experiments in animal development. For example, when the fertilized egg of a sea urchin begins to develop into the adult, it divides first into two cells, and these each divide into two more to give four. If allowed to continue development, each of these four cells will, by further division, give rise to different parts of the adult organism. But, if the four cells are broken apart and allowed to continue developing separately, they each give, not different fragments of the adult, but a complete adult sea urchin. Driesch concluded that the four cells are identical and that the normal forces of nature could not cause different parts of an adult to arise from each of four identical cells. Again, if the limbs of certain amphibia are removed, perfect new limbs develop in their place. Driesch considered that this could come about only under the influence of the directive vital force, which is attempting to resist damage to the organism, and could never occur if the component atoms and molecules of the animal were merely under the sway of the blind forces of nature of the inanimate world. The action of Driesch's vital force requires that the normal laws of nature can be violated in living organisms, and Driesch considered in detail how this violation might be detected. Whether a vital force of the kind postulated by Driesch could conceivably exist is open to philosophical argument. Whether, in fact, it does exist can be investigated by scientific experiment.

Both kinds of vitalist theory can be tested by trying to discover, as biochemists do, whether living organisms can be completely explained by the interaction of their component parts according to the normal laws of chemistry and physics. It will be seen in this book that the most fundamental processes of living cells, including their self-copying and inheritance, have now been shown to be founded on the interaction of large molecules according to the normal laws of chemistry. No new force of nature appears necessary to explain these processes and, hence, a vital force of the kind suggested by Liebig does not seem to exist. But the fundamental processes of development are not yet clearly understood. Therefore, a directive vital force of the kind suggested by Driesch cannot be excluded. Nevertheless, the spectacular discoveries in molecular biology have undermined the faith of biologists in vitalism of any kind, and evidence does, in fact, slowly accumulate against a vital force as suggested by Driesch.

In addition to these vitalist theories that relate to all living things, there are also vitalist theories about the function of the brains of higher animals. Again, the mechanist view is now dominant: that all manifestations of the brain, including the human brain, result from the interaction of its component parts according to the normal laws of chemistry and physics. If this is true, then the experience of our will being free to direct the movements of our bodies is an illusion. But there are a few philosophers and scientists who still believe that the mind may be able to direct the movements of our bodies by redirecting the interactions of the atoms and molecules of the brain to produce the required nervous impulses. Their ideas resemble those of the seventeenth century philosopher Descartes, who suggested that the human body is a machine, the behavior of which can be modified by the soul acting on the pineal gland and so influencing the nervous system.

2. The Structure of Proteins

The largest molecules of living cells are those of proteins and nucleic acids, and it will be seen later that inherited differences between living organisms are founded on differences in the structure of these molecules. The essential features of their structure will now be described.

Proteins, which we shall consider first, were isolated in the last century

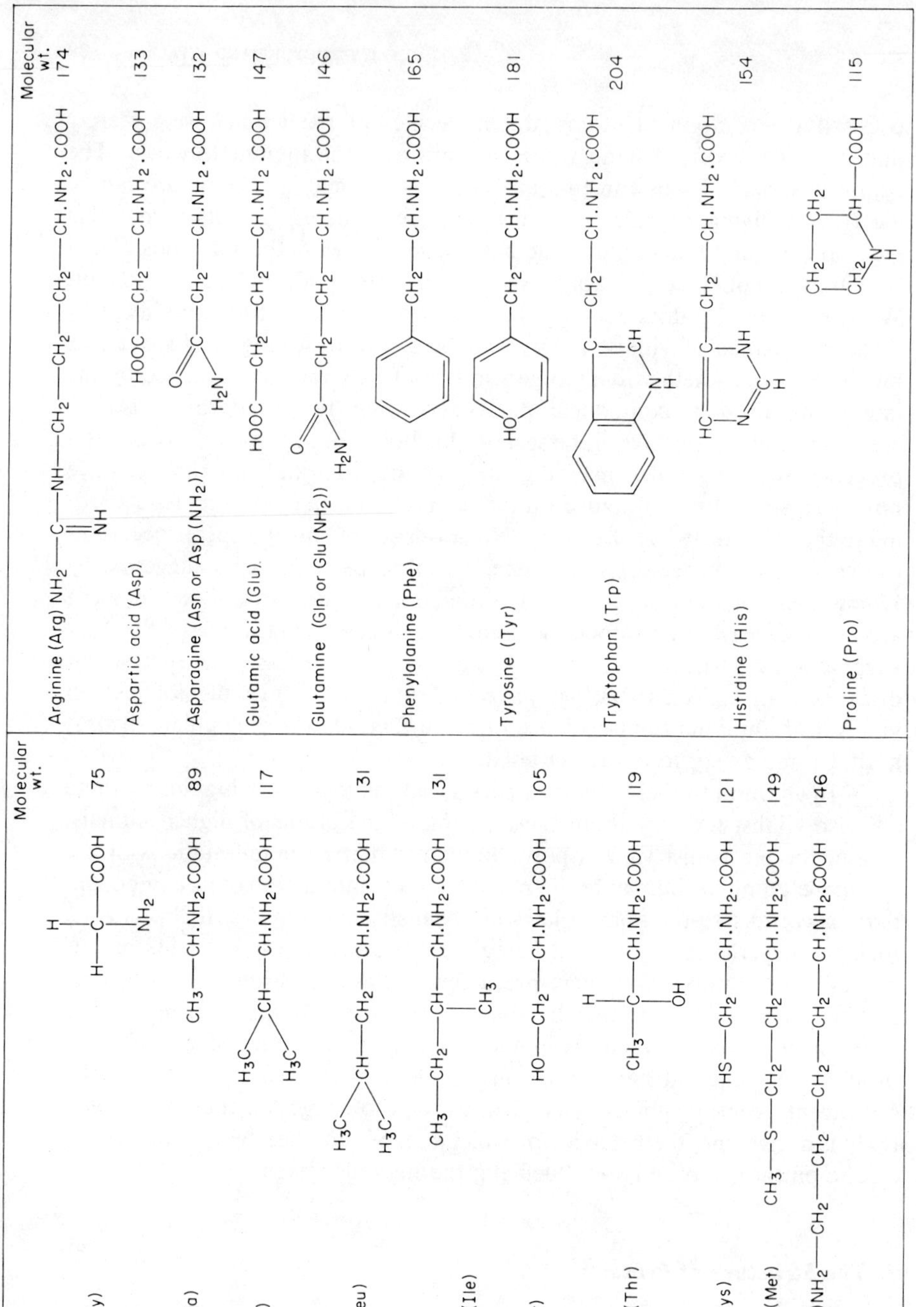

Molecular wt.

Glycine (Gly) 75

Alanine (Ala) 89

Valine (Val) 117

Leucine (Leu) 131

Isoleucine (Ile) 131

Serine (Ser) 105

Threonine (Thr) 119

Cysteine (Cys) 121

Methionine (Met) 149

Lysine (Lys) 146

Molecular wt.

Arginine (Arg) 174

Aspartic acid (Asp) 133

Asparagine (Asn or Asp (NH₂)) 132

Glutamic acid (Glu) 147

Glutamine (Gln or Glu(NH₂)) 146

Phenylalanine (Phe) 165

Tyrosine (Tyr) 181

Tryptophan (Trp) 204

Histidine (His) 154

Proline (Pro) 115

Figure 7. The 20 amino acids that occur in normal proteins and their conventional abbreviations.

from living organisms and were recognized as a group of compounds whose molecules have related, but distinct, structures. They were all found to contain the elements carbon, hydrogen, oxygen, nitrogen, and sulfur in roughly the same proportions but to differ from one another in certain properties, such as the ease with which they are precipitated with acids and salt solutions. It was soon discovered that the molecular weights of proteins are very large and that their molecules must contain some thousands of atoms. The problem of discovering the precise molecular structure of any protein appeared difficult, that is, of discovering precisely how the component atoms are linked together to form the molecule.

A start toward solving this problem was made when it was discovered that all proteins are built up from a limited number of smaller molecules called amino acids, into which they disintegrate when heated with hydrochloric acid. It is now known that 20 kinds of amino acid occur in normal proteins. Their structures are shown in Figure 7. It is seen that all except one have a carboxyl (—COOH) group and an amino (—NH$_2$) group linked to the same carbon atom. To this carbon is also linked a hydrogen atom and another organic group by which one amino acid is distinguished from the next. The structure of the one exception, proline, is closely related to that of the others. It has a carboxyl and an imino (—N—H) group linked to a common carbon atom. Some protein molecules contain all 20 kinds of amino acid, while others contain a few less.

The manner in which amino acids are linked together to form protein molecules was established around 1900 by the great German chemist Emil Fischer. He showed that the carboxyl group of one amino acid can react with the amino group of another; water is eliminated, and a *peptide* bond is formed (Figure 8). This type of reaction does not necessarily end with the joining of two amino acids, since the molecule that is formed has a free

Figure 8. Reaction of alanine and serine to form the dipeptide alanlyserine.

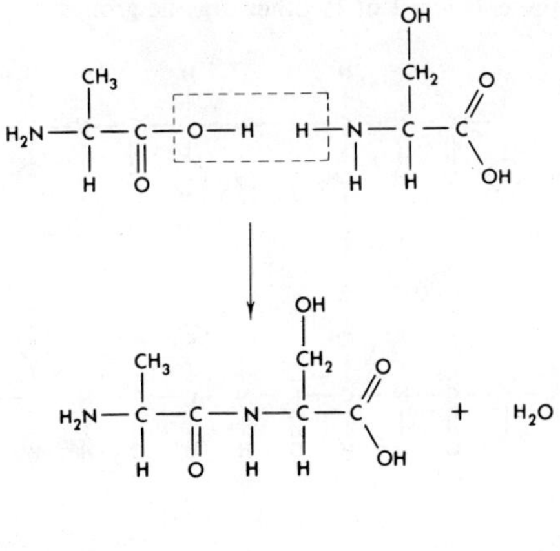

carboxyl group at one end (the *C-terminus*), and a free amino group at the other (the *N-terminus*). Therefore, either one of these groups, or both of them, can react with another amino acid. Through a series of these reactions, a long chain of amino acids can be built up. Regardless of how long this chain is, it will always have a free amino group at one end and a free carboxyl group at the other (Figure 9).

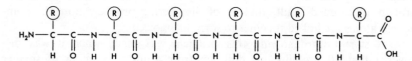

Figure 9. Peptide chain of 6 amino acids. R represents 1 of 20 possible organic groups.

From Fischer's work it became clear that protein molecules contain long chains of amino acids of this kind. Because the amino acids are joined by peptide bonds, these chains are called peptide, or polypeptide, chains. It also became clear that the molecules of some proteins contain only one polypeptide chain, while others contain two, or a few more. The chains, it was found, are linked by a reaction between residues of the amino acid cysteine on two different chains (Figure 10). (The term *residue* is used because an amino acid, when forming part of the protein molecule, has lost two hydrogen atoms and an oxygen atom and has also lost its identity as a molecule.) The molecular weights of proteins showed that each peptide chain in a protein molecule must normally be composed of 100 or more amino acid residues.

Figure 10. Two peptide chains linked by a reaction between cysteine residues. R represents any 1 of 19 other organic groups.

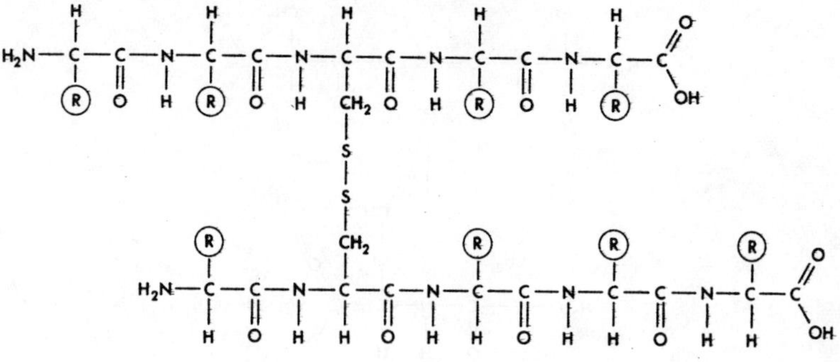

With these discoveries, the interest of chemists in the structure of proteins largely ended. By the start of the twentieth century, organic chemistry had become hardened into a clearly defined subject with set limits of interest. Proteins lay outside these limits, largely because their purity could not be demonstrated by the classic methods of organic chemistry. As a result, the study of protein structure was left to biochemists, and it was they who were able to make one of the greatest discoveries in organic chemistry of the century, the discovery of the precise arrangement of amino acids in the molecule of a protein.

The task of completely discovering the structure of a particular protein involves finding the number of peptide chains in the molecule, the number of residues of each kind of amino acid in each chain, the order in which these amino acids are arranged in each chain, and precisely how the different chains are linked. A hypothetical, and highly simplified, example may make this clear. Suppose that each molecule is found to contain two peptide chains. Chain 1 contains two residues of glycine, one of cysteine, and one of alanine. Chain 2 contains one residue of serine, one of cysteine, and one of aspartic acid. The order of the amino acids in chain 1 is found to be Gly–Ala–Cys–Gly, and that in chain 2 is Asp–Cys–Ser. (When writing the order of amino acids in a peptide chain, it is conventional to write first the amino acid that has its —NH$_2$ group left free and to write last the amino acid with its —COOH group left free.) These two peptide chains are linked by cross linkages between the cysteine residues so that the complete structure is

The word *sequence* is used to denote the description of the particular amino acid occurring in each successive position along a peptide chain. The sequence of chain 1 in the above example is thus Gly–Ala–Cys–Gly.

In the example just discussed it has been assumed that a protein can be isolated and assigned a precise structure—in other words, that a sample of a protein can be "pure" by the definition of the organic chemist, in that it is a collection of identical molecules. Protein molecules are, however, far larger than this simplified model. It might seem improbable that a living cell could repeatedly make peptide chains containing 100 or so amino acids without the slightest variation in chain length or sequence. To many biochemists around 1945, when little progress had been made in finding the precise structure of proteins, this did, in fact, seem highly improbable. They suggested, therefore, that a "pure" protein would always have slight variations in the sequence of its peptide chains from one molecule to the next; because of this variation, the problem of protein structure could never be solved precisely.

Another important group of biochemists took, at this time, a more optimistic view of protein structure—in fact, as we now know, an overly optimistic view. They suggested that the different molecules of a pure protein contain peptide chains of identical length and sequence. Moreover, they claimed that analyses of proteins suggested that a fundamental simplicity exists in their sequence, namely, that *"in every protein, each amino acid residue is distributed throughout the entire length of the peptide chain at constant intervals."* For example, it was claimed that silk protein contains residues of the amino acids tyrosine, alanine, and glycine in the ratio 1:4:8, and it was suggested that each molecule of the protein consists of a single peptide chain with the sequence

–Tyr–Gly–Ala–Gly–X–Gly–Ala–Gly–X–Gly–Ala–Gly–X–Gly–Ala–Gly–

repeated again and again over its entire length (where X is an amino acid other than tyrosine, alanine or glycine). In actual fact the ratios, found in analyses of different proteins, between the number of residues of the different kinds of amino acid were not very close to the small whole numbers that the theory required. But this, it was claimed, could be due to impurity of the proteins and inadequate analytical methods. More recent analyses of pure proteins with adequate techniques show that these simple ratios do not exist.

Experiments, which will now be described, have shown that both these theories of protein structure were incorrect. Proteins can be pure by the definition of the organic chemist and, in spite of the large size of their molecules, the sequence of the peptide chains in one molecule is identical to that in the next. The peptide chains, however, contain no repeating sequences of amino acids. These facts of protein structure are important to the correct functioning of living cells. The reason will become clearer in later chapters, but for the moment it can be illustrated by the analogy that likens a peptide chain to a word composed of amino acid "letters." If a word was never spelled in the same way twice, communication would be imprecise, while if all words had to be spelled in a rigidly repeating sequence of letters, the possible number of words would be unnecessarily limited. The precise and detailed communication between cell nucleus and cytoplasm is based on unvarying "spelling" in the peptide chains of each protein, without repeating sequences of amino acids.

Until about 25 years ago, proteins were looked on with awe by many biochemists. To regard proteins as normal organic compounds, and to attempt to discover their structure, took great courage. This was, in fact, done by A. C. Chibnall and his colleagues in the Biochemistry Department of Cambridge University, and one member of this group, F. Sanger, brought their work brilliantly to fruition by discovering, between 1945 and 1955, the arrangement of all the amino acids in the protein hormone insulin. Sanger and his colleagues chose insulin because its molecule is relatively simple and contains only 51 amino acid residues. They assumed that every molecule in a pure sample of insulin is identical to the next—an assumption that was proved to be justified by the rational results they obtained from their analyses.

Sanger's work was founded on the newly discovered technique of paper chromatography which enables amino acids and peptides to be separated from one another very simply. His work also depended on a second technique that will reveal which particular amino acid lies at the N-terminus of a peptide chain. He found that when a protein is dissolved in sodium bicarbonate solution and an alcoholic solution of 2:4 dinitrofluorobenzene is added, a molecule of this compound attaches itself to the free $-NH_2$ group at the end of each peptide chain (Figure 11). The important part of this discovery was that the compound remains attached to this group after the protein is hydrolyzed into its individual amino acids with hydrochloric acid. The resulting dinitrophenyl amino acid from the end of the peptide chain can be easily isolated because it is soluble in ether, and its particular kind of amino acid can be identified by paper chromatography. If more than one kind of dinitrophenyl amino acid is obtained, then the protein must have more than one amino acid chain per molecule.

Sanger treated insulin with this reagent and then hydrolyzed it into its component amino acids. He found that the dinitrophenyl derivatives of glycine and phenylalanine were formed in the amounts expected if insulin has two peptide chains, one beginning at the N-terminus with glycine and the other with phenylalanine. By treating insulin with an oxidizing agent, Sanger was able to break the linkage between the two peptides and isolate each separately. He then set about to discover the sequence of each.

If a protein is heated with hydrochloric acid to 37°C instead of to boiling, it is not hydrolyzed completely into its component amino acids. The peptide bonds are broken slowly, and a mixture of an enormous number of peptide fragments is formed. But since certain peptide links in the protein are broken more rapidly than others, certain fragments of the protein will predominate over the others. Sanger warmed each of the peptide chains of insulin in this way and was able to isolate certain peptide fragments chemi-

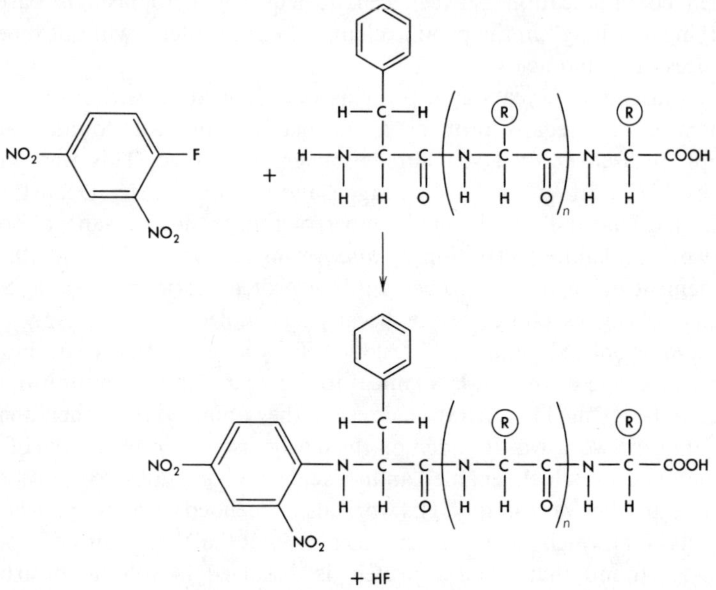

Figure II. Sanger's reagent, 2:4 dinitrofluorobenzene, reacting with phenyl-alanine residue at end of protein chain.

cally pure. He set about finding the sequence of these peptides in the following way. If a peptide contained only two amino acids, he was able to establish its sequence immediately by reacting it with 2:4 dinitrofluorobenzene: the amino acid that reacted with this reagent was the one at the N-terminus. If a peptide contained three amino acids, the one at the N-terminus could again be identified, but the order of the other two remained unknown. The tripeptide was then warmed with hydrochloric acid, and one of the two possible dipeptides was isolated. Its sequence was determined, and from this the sequence of the tripeptide could be deduced. The sequence of larger peptides were found by an extension of this method.

From the sequence of overlapping fragments of the peptide chains of insulin, Sanger was able to deduce the sequence of each chain. For example, the chain beginning in glycine yielded the fragments Ser–Leu–Tyr and Leu–Tyr–Glu. Sanger concluded that they were derived from the following sequence in the peptide of insulin: Ser–Leu–Tyr–Glu. The principle is the same as if we knew that a word of seven letters contained the sequences INS, SULI, and LIN. The word is clearly INSULIN.

Sanger had isolated the two peptide chains of insulin by first oxidizing the protein and so breaking the linkages between one residue of the amino

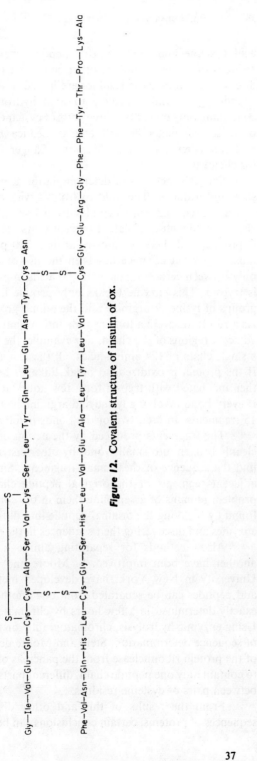

Figure 12. Covalent structure of insulin of ox.

37

acid cysteine and another. One peptide chain of insulin contained four residues of cysteine and the other contained two. It was now necessary to discover which of these residues are linked into pairs. He discovered this by submitting the intact protein to mild hydrolysis and isolating fragments containing only one of the three pairs of cysteine residues. From the sequences of these fragments he was able to deduce that of the complete insulin molecule (Figure 12). For this work, Sanger was awarded the Nobel Prize for chemistry.

Sanger's methods of determining the sequences of peptide chains were slow and arduous. The mild hydrolysis with acid gives a complex mixture of fragments, and pure peptides from overlapping regions of the original chain are difficult to isolate. In recent years the techniques have been greatly improved, and the structures of around 100 proteins are now known. The most important advance has been the use of enzymes that break linkages only between certain kinds of amino acids in protein chains; the most useful is trypsin. This enzyme breaks only peptide linkages between the carboxyl groups of lysine or arginine and the amino groups of any other amino acid. As a result, it severs a long peptide into a limited number of fragments from adjacent regions of the chain. For example, the protein ribonuclease contains a single chain of 124 amino acids. Of these, 4 are arginine and 10 are lysine. If the protein is oxidized, to break linkages between cysteine residues, and then incubated with trypsin for a few hours, it is broken into 13 fragments. If every bond involving lysine and arginine had been broken, there would be 15 fragments. In fact, two of the bonds involving lysine are not susceptible.

The fragments produced by the action of trypsin can, if necessary, be cleanly broken into smaller ones by other enzymes. It is then not difficult to find the sequence of these small fragments. Since these fragments are from adjacent segments of the original peptide chain and do not overlap, the problem remains of assembling them in the correct order. This order can be found by breaking the original peptide into different fragments with different enzymes and discovering their sequences in the regions of lysine and arginine.

Also, methods for separating amino acids and peptides from one another have been improved. S. Moore and W. H. Stein, of Rockefeller University in New York, have developed methods by which amino acids and peptides can be separated from one another and the quantities of each exactly determined in a few hours by chromatography on columns of resin. Using enzymic hydrolysis, chromatography on resins, and improved methods of sequence determination, Stein and Moore deduced the complete sequence of the protein ribonuclease from the pancreas of the ox (Figure 13). It is seen to contain only one peptide chain, different parts of which are linked by bonds between pairs of cysteine residues.

From the results of this and other discoveries of the amino acid sequences of proteins, certain conclusions can be drawn about their structure.

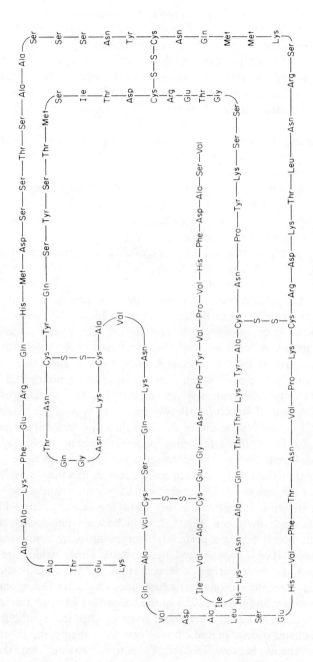

Figure 13. Covalent structure of pancreatic ribonuclease of ox.

It is clear that in a sample of a pure protein every molecule is, in fact, identical to the next. Surprising as it may seem, there is no variation in sequence from one molecule to the next. Moreover, along each peptide chain the amino acids appear to be in a complete jumble; that is, a certain amino acid does not appear at regular intervals along the chain, nor do two amino acids appear to occur next to one another more often than would be expected by chance. Most proteins differ completely in their sequences, but that of closely related proteins can be very similar. Sanger, for example, studied, as well as ox insulin, insulin from pig, sheep, horse, and whale. The structure of each is identical except for a sequence of three amino acids, which differs from one species to the next.

From what has been said so far it might be thought that the peptide chains of proteins are free to indulge in snakelike movements and can assume a large number of conformations. In fact, it appears that, even in solution, they cannot. The molecules of most pure proteins, as well as having identical amino acid chains, are apparently coiled into structures of identical shape. This three-dimensional structure has recently been discovered in great detail for a number of proteins by the physical method of X-ray diffraction. X-ray diffraction patterns were first obtained from protein crystals many years ago, but the problem of deducing the structure of the molecules was too complex to be solved. But, in recent years an important advance in technique has simplified the calculation of the results to a level that can be handled by fast electronic computers. Using these methods, M. F. Perutz and J. C. Kendrew and their colleagues at Cambridge University, after thousands of measurements and calculations, deduced in great detail the three-dimensional structures of two proteins: hemoglobin from blood and myoglobin from muscle. Since then the three-dimensional structures of a number of other proteins including pancreatic ribonuclease (Figure 14) have been deduced. Physical measurements on solutions of these proteins show that they maintain almost identical structures in solution, unless they are disrupted (or *denatured*) by adverse conditions such as high or low pH. These three-dimensional structures are held together by three kinds of noncovalent bond: hydrogen bonds, largely between imino and carbonyl groups; ionic bonds between positively charged nitrogen atoms and negatively charged carboxyl groups; and hydrophobic bonds, which are attractions between the hydrocarbon side chains of certain amino acids. The characteristic three-dimensional structure appears to be the spontaneous result of a protein being synthesized in the cell with a particular amino acid sequence (see p. 113). This has been confirmed by chemical synthesis of the polypeptide chains of insulin and ribonuclease from their component amino acids: these chains became biologically active, proving that they had spontaneously folded into the correct three-dimensional protein structures as synthesis proceeded. Some protein molecules are composed of a small number

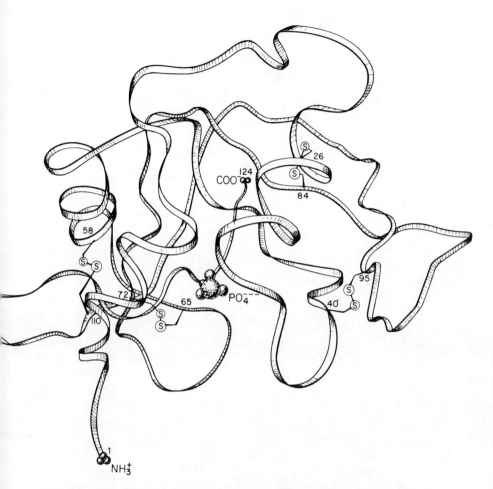

Figure 14. Conformation of pancreatic ribonuclease of ox. The N-terminus (amino acid 1) and C-terminus (amino acid 124) and the cysteine residues that form disulphide bridges have been numbered. The site of enzyme activity is where a phosphate ion is shown to be bound. (By kind permission of Dr. D. Harker, Dr. E. Kartha, and John C. Wallace.)

of polypeptide chains held together by the same noncovalent bonds that hold the polypeptide chains themselves into three-dimensional structures. An example is the human hemoglobin molecule, which is composed of two identical α chains and two identical β chains. As will be seen, these same weak bonds also appear to be responsible for the binding of molecules into larger cell structures.

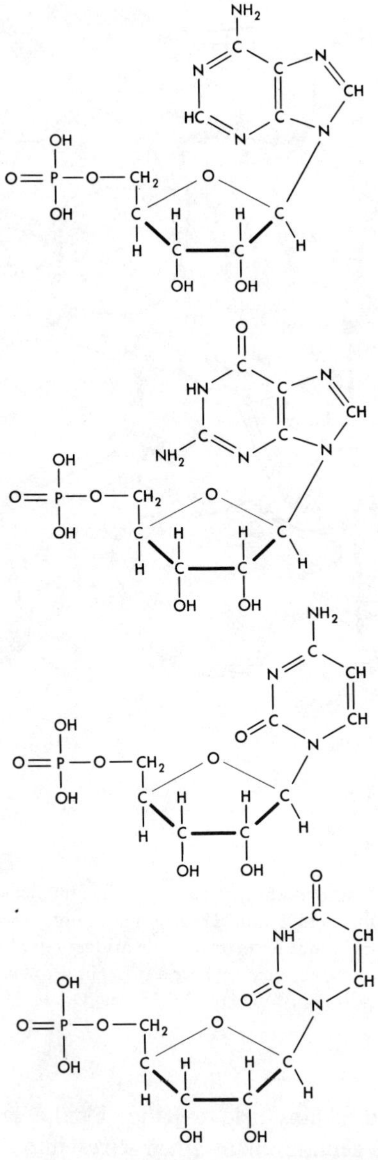

Adenylic acid
(base adenine)

Guanylic acid
(base guanine)

Cytidylic acid
(base cytosine)

Uridylic acid
(base uracil)

Figure 15. Four nucleotides of RNA. In the center of each nucleotide is the ring of the sugar ribose containing four carbon atoms and one oxygen atom.

3. The Structure of Nucleic Acids

Nucleic acids are of two distinct types: ribonucleic acids (abbreviated to RNA) and deoxyribonucleic acids (abbreviated to DNA). Their molecules are very large, but, as with proteins, the problem of discovering their structure is simplified by the fact that they can be hydrolyzed into smaller molecules called nucleotides. This disintegration of nucleic acids into nucleotides reminds us of the disintegration of proteins into amino acids, and, in fact, nucleic acid molecules are long chains of linked nucleotide residues. But whereas a protein molecule contains up to 20 kinds of amino acid, many nucleic acids contain only 4 kinds of nucleotide.

The four nucleotides found in all RNA are shown in Figure 15. It is seen that they all contain the sugar ribose, and all have a residue of phosphoric acid linked to carbon 5 of this ribose by an ester link. They differ in the molecule that is attached to carbon 1. This is one of four closely related organic compounds named adenine, guanine, cytosine, and uracil. They are called bases because they contain basic nitrogen atoms. In many RNAs the bases are modified by the addition of other groups, such as methyl groups.

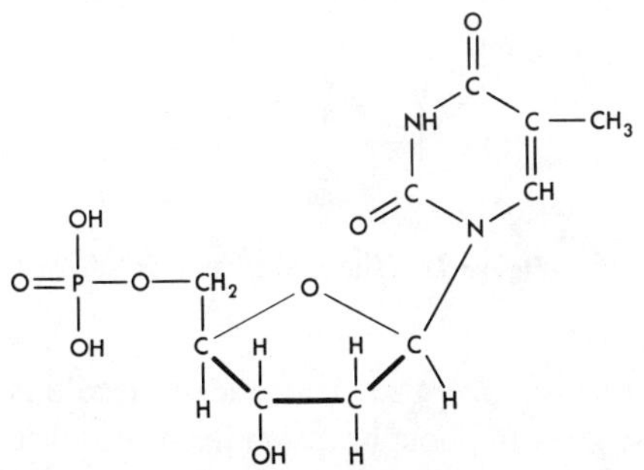

Figure 16. Thymidylic acid, one of the four nucleotides of DNA.

The four nucleotides of DNA closely resemble those of RNA but differ in that the sugar has no oxygen atom on carbon 2; hence it is named 2-deoxyribose. Also, one of the four bases is different: uracil is not found in DNA, but the closely related base thymine is found instead. Moreover, the DNA of certain bacteriophages has 5-hydroxymethylcytosine instead of cytosine. The structure of thymidylic acid, the deoxyribonucleotide containing thymine, is shown in Figure 16.

In RNA and DNA the nucleotides are linked as shown in Figure 17.

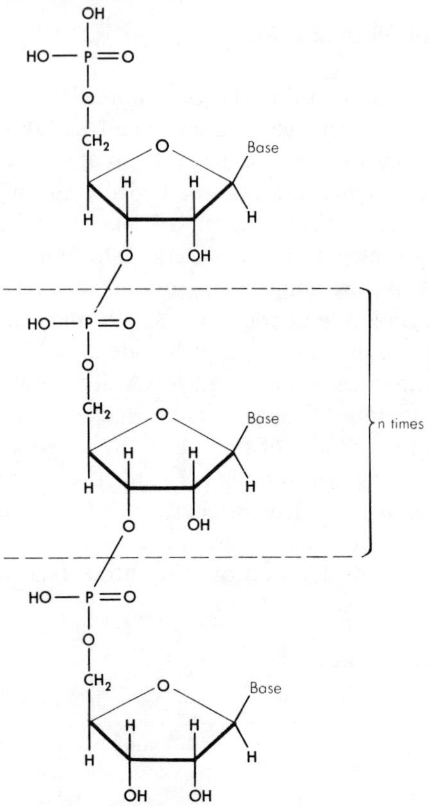

Figure 17. The structure of RNA.

The simplest way to visualize a nucleic acid molecule is as a chain of alternating residues of sugar and phosphoric acid, a molecule of a base being linked to carbon 1 of each sugar residue. Just as a protein chain has a N-terminus and a C-terminus, so the ends of a nucleic acid molecule are not identical: one end has a sugar residue with its C–3 not linked to another nucleotide, and the other end a sugar residue with its C–5 not so linked. These are the 3' and 5' ends, the prime marks indicating numbers of sugar carbon atoms as opposed to atoms of the bases. The sugar phosphate chain is the same in every RNA or every DNA molecule, but one molecule can be distinguished from another by the length of this chain and by the sequence of bases along it. At first sight it might seem that because there are only four kinds of nucleotide in DNA and in many RNA molecules the number of possible molecules of each would be severely limited. This is not so: many billion different chains of 100 nucleotides could be built up from four kinds of nucleotide.

As with proteins, the question arises, Can a sample of a nucleic acid be isolated that is pure by the definition of the organic chemist in that it is a collection of identical molecules? The answer again is yes. Samples of RNA have been isolated from the cytoplasm of cells whose complete nucleotide sequences have been determined. As with proteins, the consistent results that were obtained show that the samples were pure. These RNAs are the transfer RNAs, which, as will be seen later, transfer amino acids into peptide chains during protein synthesis. The first to be isolated pure was a transfer RNA for the amino acid alanine, and its nucleotide sequence was determined by

Figure 18. Covalent structure of alanine transfer RNA. The letter *p* denotes internucleotide phosphate. When placed to the left of the capital letter of a base it represents the 5′-phosphate of the nucleoside bearing that base; when placed to the right it represents the 3′-phosphate. Often the letter *p* is replaced by a hyphen with the same significance. The molecule has been drawn in the probable "cloverleaf" folding which is stabilized by hydrogen bonds between pairs of bases (see p. 72).

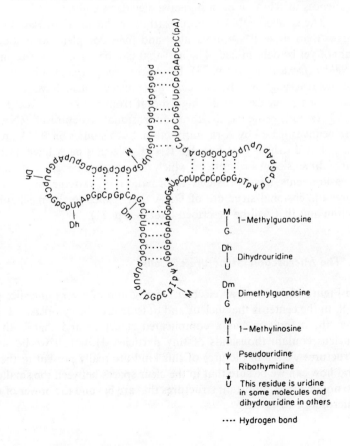

M — G	1–Methylguanosine
Dh — U	Dihydrouridine
Dm — G	Dimethylguanosine
M — I	1–Methylinosine
ψ	Pseudouridine
T	Ribothymidine
*Ů	This residue is uridine in some molecules and dihydrouridine in others
····	Hydrogen bond

R. W. Holley and his colleagues at Cornell University around 1964. Their techniques were similar to those used to determine the amino acid sequences of proteins. By digestion with two enzymes, ox pancreatic ribonuclease and T_1 ribonuclease from yeast, they broke the RNA molecule of 77 nucleotides into smaller fragments whose nucleotide sequences were determined. Sequences of fragments that overlapped allowed the sequence of the complete molecule to be deduced (Figure 18). It is seen to contain a number of less common bases.

In this work Holley separated the nucleotide fragments by column chromatography, and for his analyses he needed many milligrams of RNA. Sanger has since developed methods by which the nucleotide sequence in a few micrograms of transfer RNA can be completely determined in only a few months. The organism from which the RNA is isolated is first made highly radioactive with ^{32}P—and this can be done only on microorganisms, which is a limitation of the method. The fragments from enzyme digestion of the RNA are then separated by paper chromatography and detected by their radioactivity. This technique is now being applied to determining nucleotide sequences in RNA of even higher molecular weight.

The smallest DNA molecules (those of bacterial viruses) are very much larger than these RNA molecules, and their complete nucleotide sequences cannot yet be determined. The best that can be done is to determine *nearest-neighbor frequencies* (see p. 78). DNA from each organism has its characteristic nearest-neighbor frequencies and hence must have a characteristic nucleotide sequence that distinguishes it from the DNA of other organisms.

As with proteins, the three-dimensional structure of RNA and DNA can be investigated by X-ray diffraction. Such studies on RNA are at an early stage but do show that transfer RNA molecules each have a characteristic three-dimensional structure on which their activity depends. X-ray diffraction measurements have provided information of extreme importance on the three-dimensional structure of DNA that relates to the mechanism of gene action and that will be described later (see p. 71).

4. The Microscopic Structure of Cells

In Figure 19 is shown an electron micrograph of a very fine slice of pancreas cell. In the center is the nucleus and outside it is the cytoplasm. It seems clear that the cytoplasm has a complicated structure and that both it and the nucleus contain thousands of tiny particles. How can we be sure that the structures we see in pictures of this kind are really present in the living cell; and how can we be sure that in the clear spaces between the smallest particles do not lie other important structures that are beyond the power of the electron microscope to reveal?

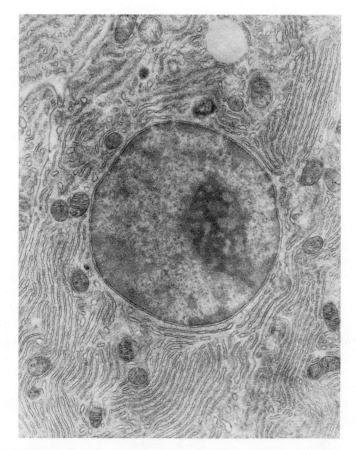

Figure 19. Electron micrograph of section through the center of a cell from the pancreas of a bat, magnification approximately 10,000 diameters. (Courtesy Dr. D. W. Fawcett.)

Before viewing in the electron microscope, cells must be killed by being treated rather violently with certain chemicals—a process known as fixing. They are often treated with a compound of the metal osmium and then dried and sliced very thinly. The osmium atoms have varying affinities for different structures in the cell and become attached to them to different extents. The resulting electron micrographs are largely pictures of the distribution of osmium in the cell slice, since the cell structures themselves have little effect on the electron beam. Do the structures seen in such pictures all exist in the living cell, or are they partly spurious patterns produced by the harsh treatment?

The problem is by no means a new one in microscopy. For many years, fixation and staining have been used to reveal cell structures in the light microscope and the question has repeatedly arisen as to whether some of these are artifacts produced by the treatment. In these instances, numerous experiments have proved that staining has very largely revealed true cell structures. Experiments are also proving that most of the structures seen in the electron microscope are genuine cell components. For example, tiny particles can be isolated, by methods which will be described, from the cytoplasm of cells that have not been fixed. Their size and shape, as measured in the ultracentrifuge, are those of the smallest particles that can be seen in the cytoplasm of cells under the electron microscope.

These particles have been named ribosomes, for reasons that will become apparent later. We now come to the second question. How can we be sure that in the clear spaces between the ribosomes do not lie other important structures that the electron microscope is unable to reveal? The answer comes first from experiments that show that each ribosome has about 4 million times the weight of a hydrogen atom. This can be roughly calculated from their size in electron micrographs, and calculated more accurately from their speed of sedimentation in the ultracentrifuge. A medium-sized protein or nucleic acid molecule has a molecular weight of around 40,000; hence a ribosome could contain only around 100 of these molecules. Therefore, any particles in the cytoplasm, invisible in the electron microscope, must have considerably less than 100 times the size of an average protein or nucleic acid molecule. Second, when cell particles are isolated by centrifuging broken cells, in a manner that will soon be described, it can be shown clearly that structures in the cytoplasm smaller than ribosomes are, in fact, individual molecules.

An electron micrograph, such as that in Figure 19, therefore reveals the smallest organized structures of a cell. It will be seen in later chapters that the basis of inheritance lies in the interaction of large molecules with these structures. This is to be expected. If we tried to discover how a cell functions by studying its small molecules, we would have reached too simple a level of organization—like a man who tried to discover how a car works by making a chemical analysis of its valves and pistons. An important point follows from this fact. Physicists have demonstrated that very small objects, of the size of an electron, do not behave like the solid objects around us in everyday life but as a strange blend of solid object and wave, and the laws of nature prevent us from ever gaining enough precise information about them to predict accurately their future behavior. Actually, these properties are possessed to an infinitesimal degree by the objects of everyday life, but they have no practical importance. Neither have they any practical importance with objects the size of protein and nucleic acid molecules, and hence these properties need not be considered in explaining the basis of inheritance.

Many kinds of cells have been viewed under the electron microscope, and although they vary in size and shape, it is quite clear that they have many structures in common. The structures that it has been concluded, exist in a "typical" animal cell are shown in the diagram of Figure 20. In the center is the large, roughly spherical, nucleus. The nucleus is bounded by a membrane, and the bulk of the material within it appears as a mass of coarse particles. The particles within the nucleus have been given the noncommittal name of chromatin because they are readily stained with certain dyes and can then also be clearly seen in the light microscope. Also, within the nucleus lies one small sphere, or a few small spheres, called the nucleoli. They also contain granules. The nucleus has this appearance during most of the life of the cell, but before cell division it changes radically, as will be described later.

Figure 20. Diagram of typical animal cell as seen under the electron microscope.

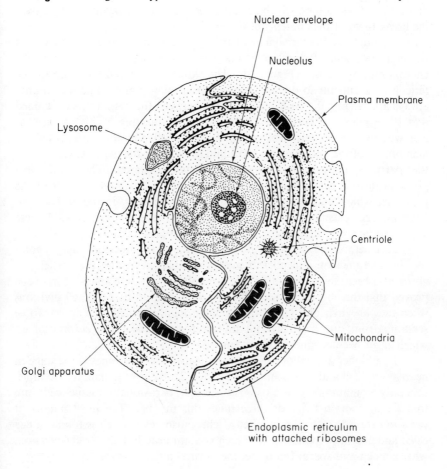

The largest structures normally found outside the nucleus, in the cytoplasm, are the mitochondria, which are shown in the diagram of Figure 20. They have a fairly complicated internal structure and are important, but we shall not be concerned with them in this book. They contain enzymes whose function is to oxidize acetic acid, which is produced from more complex compounds, to carbon dioxide and water, with the production of energy. Thus they are power units and are found in large numbers in cells that need large amounts of energy, such as muscle cells. Bacteria do not have mitochondria, but they usually perform the same chemical reactions, with similar enzymes differently arranged. Plant cells contain mitochondria and, in addition, often contain chloroplasts that are of similar size and somewhat related structure. Their function is to absorb light energy and store it in compounds that provide chemical energy to the plant. Other particles found in the cytoplasm that need not concern us are centrosomes, involved in cell division, and lysosomes, which are capsules containing enzymes that catalyze the breakdown of the contents of the cell.

Throughout the remainder of the cytoplasm of most cells, other than bacteria, a system of membranes branches out, which has been named the endoplasmic reticulum (Figure 21). The three-dimensional structure of the reticulum is difficult to deduce from electron micrographs. The membranes always appear in pairs with a narrow gap between them, and it is concluded that these gaps are part of a system of interconnecting canals—somewhat like woodworm tunnels in a rotten wooden ball. On the surfaces of the membranes of the endoplasmic reticulum away from the gaps between them, the particles named ribosomes are always scattered. They are most plentiful in cells that are rapidly forming proteins, such as the cells of the pancreas, which make digestive enzymes. Certain cells, including bacteria, have no endoplasmic reticulum, but they do have ribosomes, and these appear to be freely distributed through the cytoplasm.

Just before cell division, the appearance of the nucleus changes completely. The nucleoli and nuclear membrane disappear, and the particles of chromatin become concentrated into a number of chromosomes. It has been proved that the chromosomes still remain intact even between cell divisions when they are invisible. The particles of chromatin, which then appear to be separate from one another, still in fact form part of the chromosomes, which have become finely drawn out.

In a higher animal or plant the number of chromosomes is the same in nearly every cell and is typical of the species. The cells of a man, for example, generally contain 46 chromosomes, though occasionally tissue cells are found that contain twice or three times this number. The most important variation from the normal number of chromosomes per cell is found in egg cells, and sperm or pollen cells, which contain exactly half. At fertilization, when an egg and sperm cell unite, the normal number is restored.

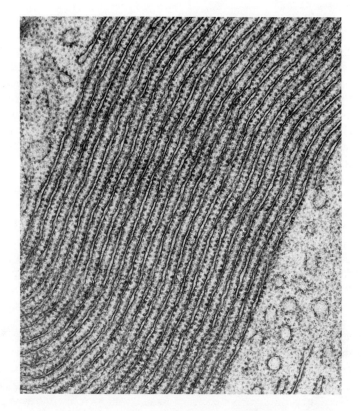

Figure 21. Electron micrograph of endoplasmic reticulum showing attached ribosomes, magnification 30,000 diameters. (Courtesy of Dr. D. W. Fawcett.)

5. The Isolation and Analysis of Cell Structures

The electron microscope reveals many structures in cells but gives little clue as to what they are made of. The most direct way to find this is to isolate and analyze each kind of structure. For many years biochemists tried to do this by grinding up cells in salt solutions and centrifuging the resulting suspension first at low and then at high speeds. They hoped that first the larger and then the smaller particles would sediment. But they could see, by examining the various sediments under the microscope, that they were getting no clear-cut isolation of the different cell structures. Then, in the 1940s it was discovered that the reason for this failure was that many of the cell structures burst when they entered the salt solution and that this bursting could be prevented by substituting a solution of cane sugar. Successful methods of isolating cell structures were then soon developed.

The methods that are used will be illustrated by describing a particular isolation. Suppose that we wanted to isolate various structures from the cells of liver. The procedure, like many scientific operations, does not require extreme skill, although to devise it took several years of inspired trial and error in the laboratory. We would first prepare, and cool in a refrigerator, a 0.25M sucrose solution containing 1.5 mM of magnesium chloride which is necessary to prevent cell structures from disintegrating and clumping. If we decided to use rat's liver, we would kill a rat and rapidly dissect its liver. Then we would weigh out a few grams of it and drop it into the recommended volume of the ice-cold sucrose solution. Next, in a refrigerated room, we would disintegrate the liver in a Waring blender, the rotating blade of which sets up shearing forces in the liquid that tear the cells apart and liberate their contents.

We would then put the resulting suspension in a small centrifuge tube and centrifuge it in a refrigerated centrifuge. By spinning the tube for 10 minutes at a slow speed that will exert 700 times the force of gravity on the suspended particles, we would get a sediment that is largely cell nuclei but that also contains unbroken cells and fragments of cells. To get purer nuclei, we would pour off the liquid from above the sediment and resuspend it in more sucrose solution and centrifuge once more.

The liquid we first poured off from above the sedimented nuclei contains the smaller cell structures. If we spin this in another tube at a speed that exerts 5,000 times the force of gravity, the sediment that forms is largely mitochondria. They can again be purified by recentrifuging. To isolate smaller cell structures, it is necessary to use a preparative ultracentrifuge. This is no larger than other preparative centrifuges but is made with great precision and is expensive. It can rotate plastic tubes at up to 70,000 revolutions per minute, and the tubes are spun in a vacuum to prevent heat being generated by friction.

If we spun the liquid remaining from the isolation of the mitochondria for 1 hour at a speed that gives 100,000 times the force of gravity, we would obtain a sediment of particles that were originally named microsomes. This name is unfortunate, for the electron microscope has since shown that these are fragments of the endoplasmic reticulum and not worthy of a special name. Attached to their surfaces can be seen ribosomes, and it has been found that if they are stirred with a solution of a detergent, the endoplasmic reticulum dissolves, leaving a suspension of ribosomes. These can then be isolated by spinning once more at 100,000 times the force of gravity. No matter how long we centrifuge, no more particles sediment out of the liquid that remains after separating the microsomes. In an analytical ultracentrifuge it can be seen that, at 100,000 times the force of gravity, individual protein and nucleic acid molecules begin to migrate through this liquid and that it contains no larger particles. Hence the cells contain no structures smaller than ribosomes —proof of the point that was discussed earlier.

If we isolate cell structures by these methods, we find that the nuclei, mitochondria, and ribosomes, and also the *supernatant* fraction that remains when these have been removed, all contain a substantial proportion of proteins. Apart from water, proteins make up quite the largest part of the mitochondria and supernatant fraction, and it is probable that virtually all their proteins are enzymes. The greater part of the nonaqueous portion of nuclei is also composed of proteins. These are partly enzymes, but partly peculiar proteins called histones, which are small molecules containing a high proportion of amino acids with basic side chains. Chromatin (the material of chromosomes) can be isolated by disintegrating the nuclei and centrifuging at a low speed and is found to consist of up to 50% of histones and to contain all the histones of the nucleus. Proteins also make up about 45% of the dry weight of ribosomes.

RNA is also widely distributed through the cell. Ribosomes, as their name denotes, contain a high percentage (about 55% of their weight); hence they are composed solely of RNA and protein. As a result, a large part of the RNA of a cell is concentrated in the ribosomes. In liver cells, ribosomes contain about one-half of the total RNA. Nuclei and the supernatant fraction contain only a small percentage of RNA. Most of the RNA of the nuclei is concentrated in the nucleoli, while in the supernatant, the RNA consists of individual molecules in solution. Although the percentage of RNA in these fractions is low, because they make up a large proportion of the total weight of the cell, they contain a fairly large proportion of the cell's RNA. About 12% of the RNA of liver cells is contained in the nuclei and about 34% in the supernatant fraction. Mitochrondria contain only a small percentage of RNA and only a small proportion of the RNA of the cell. The important concentrations of RNA within cells are, therefore, in the ribosomes, in the nuclei, and in the water of the cytoplasm. It will be seen later that RNA in each of these three places plays an important role in gene expression.

When animal cell fractions are analyzed for DNA, it is found to be almost entirely confined to the nucleus. It can also be shown, by staining cells with stains that react only with DNA, that it is also absent from every part of the nucleus except the chromatin of a nondividing cell and the chromosomes of a dividing cell. If chromosomes are isolated, they are found to contain up to 40% DNA. Mitochondria also contain small amounts of DNA, as do chloroplasts in plant cells.

DNA thus has a remarkably restricted distribution in cells. There is another peculiarity that marks DNA off from the other components of living organisms, which is revealed by measuring the total quantity of DNA in different cells. Suppose that we want to determine the quantity of DNA per cell in different tissues of a hen. One way to go about this is to kill a hen and isolate samples of nuclei from its liver, kidney, spleen, and other tissues. These are suspended in a solution of sucrose. A known fraction of the suspension is then taken, and its DNA content is determined, and the number of nuclei in another known fraction is counted under the microscope. It is then possible to calculate the average quantity of DNA per nucleus, and hence per cell, from which the nuclei came. This experiment has been done, and the results published. The average quantities of DNA per liver, kidney, and spleen cell of the hen were found to be 2.6, 2.3, and 2.6×10^{-12} gm, respectively. The amount in a red blood cell of hen was also determined on a suspension of these cells and found to be 2.6×10^{-12} gm. Values very close to these have been found on other domestic fowls, and it appears that the average quantity of DNA per cell is, at least very nearly, constant. An important exception to this rule was discovered. The average amount of DNA per cell in a suspension of sperm cells from a cockerel has been found to be 1.3×10^{-12} gm, or just half the quantity in other body cells. Similar experiments have been done on other species. Again, the quantity of DNA per body cell is very nearly constant in any one species but differs between species, and again the sperm cells contain one half of the DNA in tissue cells. For example, the tissue cells of cow or steer all contain very nearly 6.0×10^{-12} gm of DNA per cell, while the sperms of bull contain 3.0×10^{-12} gm.

These methods determine the average quantity of DNA per cell in a sample of cells. With an apparatus known as a quantitative cytophotometer, it is possible to measure the quantity of DNA in a single cell. In this apparatus, a section of a tissue, or a smear of separate cells, is placed on the slide of a special quartz microscope, and a beam of ultraviolet light is focused onto the nucleus of a single cell. The light that passes through the nucleus is allowed to fall on a photoelectric cell, and its intensity is compared with that of a similar beam that has merely passed through the quartz slide. The difference in intensity depends on the quantity of DNA in the nucleus, which can be calculated. This method again shows that the quantity of DNA in every single cell of any one species is at least very nearly the same, with a few exceptions. These exceptions are that the sperm cells, and also the egg cells, contain half the quantity in normal cells and that tissue cells, which have two or three times the normal number of chromosomes, contain two or three times the normal amount of DNA. From all these facts an important conclusion may be drawn: for a single species, a complete set of chromosomes always contains the same quantity of DNA.

Chapter 3

The Molecular Structure
of Genes

I. Bacterial Transformation Shows that Bacterial Genes Are
Made of DNA

We have seen that DNA has a peculiar distribution in the cells of higher organisms: apart from small quantities in mitochondria and chloroplasts it is confined to the chromosomes of the nucleus. Moreover, the quantity of DNA in every nucleus is, with few exceptions, constant and characteristic of the species. One important exception is that the sperm nucleus, which has half the number of chromosomes of nuclei of body tissues, also has half the quantity of DNA. These facts mark off DNA from other cell components and suggest that it plays an important part in the function of chromosomes, and hence of genes. In this chapter we shall in fact see that genes of all organisms are made solely of DNA—with the exception of a few organisms in which they are made of RNA.

Experiments that provided the first clear evidence that genes of bacteria are made of DNA will now be described. This evidence came from a study of bacterial transformation that was done by F. Griffith in England in 1927. Pneumonia bacteria, or pneumococci, exist in a number of strains or races. For the most part they have a carbohydrate coating, or capsule, surrounding the cell, but some have no capsule. When grown under identical conditions, the bacteria with capsules produce more bacteria with capsules, while those without produce more bacteria without. The difference in capsule must therefore be a genetic difference, that is, a difference in the form displayed by one or more unit characters.

Griffith was working with two strains of pneumococci: one with capsules that, when injected into mice, multiplied and killed them with pneumococcal infection, and another without capsules that multiplied but did not kill the mice. When bacteria with capsules were heated to 60°C, they were killed and no longer infected the mice. But Griffith made a strange discovery. If these dead bacteria were injected together with the live harmless bacteria without capsules, the mice sometimes died of infection, and the blood of these animals was always infested with live harmful bacteria with capsules. These bacteria could be isolated and were indistinguishable from the bacteria that had been killed before injection. It must be concluded that something passed from the dead bacteria into the live ones that caused their progeny to possess an alternative form of one or more unit characters: namely to possess capsules and be pathogenic to mice. At least at first sight, it appears that one or more genes must have passed from the dead to the live bacteria.

To try to discover how this bacterial "transformation" took place, other biologists took over where Griffith had left off. They first managed to reproduce the transformation of the pneumococci in a broth culture. Dead bacteria with capsules were added to the broth, which was inoculated with live bacteria without capsules, and then incubated for some hours. A drop of the culture was then diluted and smeared over agar jelly, which contained bacterial nutrients, in flat dishes. (This is a standard technique to deposit separate bacterial cells over the agar. After incubation, separate colonies appear, each of which is derived from a single cell.) It was found that the bacteria in most of the colonies that formed were without capsules. But about 1 in 1,000 did have capsules and hence had been transformed.

The next logical step was to extract various chemical components from the dead bacteria and see whether any of these would cause transformation. Experiments of this kind were largely done by O. T. Avery and R. D. Hotchkiss of Rockefeller University. They were, in fact, able to obtain extracts that would cause transformation, and they fractionated these extracts into different chemical components and tested each in turn. After some years of work, they announced that one, and only one, type of chemical compound from the capsulated bacteria would transform some of those without capsules when added to the broth culture. This compound was DNA that was about 95% pure. Warnings were at once issued by some biologists that the transformation might be caused by some more subtle "principle" in the 5% impurity. These warnings were justified since many thought DNA to be incapable of existing in as many structural variations as there are genes in an organism, but there was also a reluctance to finally attribute a genetic change to a chemical compound. However, the evidence later became overwhelming that the transformation is caused by the bacteria that do not have capsules absorbing DNA from those that do. For example, the DNA was further purified until it contained negligible amounts of other compounds, including less than 0.02% protein. Also, the potency of the DNA was rapidly destroyed by enzymes that break down DNA but that is unaffected by those that break down RNA or proteins.

How were these experiments to be interpreted? The progeny of the transformed bacteria possessed unit characters in forms in which their ancestors did not possess them. Therefore, either the transforming DNA contained genes responsible for these forms of the unit characters, or it activated genes that lay latent in the recipient cell. It appeared certain that the DNA, in fact, contained the genes. Transforming DNA of many different kinds could be isolated, each specific for different unit characters. For instance, if DNA from cells that are resistant to the drug streptomycin was added to a culture of those that are sensitive, some bacteria resulted that bred resistant progeny. There was no reason to believe that transforming DNA that was specific for any form of any unit character could not be isolated. Moreover, cells were sometimes transformed for two or more characters, and the fact that linkage maps could be constructed from the differing frequencies with which different pairs of genes are transferred suggested that the transformed cell absorbs chromosome fragments (see p. 23). The hypothesis that transforming DNA merely activates preexisting genes in the recipient cell, therefore, crumbles, since each gene would have to have a specific activator that is reproduced generation after generation. It is difficult to escape the conclusion that in pneumococci, and in a number of other bacteria in which transformation has been demonstrated, genes are made of DNA.

It was at first difficult to conceive how large DNA molecules could be absorbed by bacteria and become incorporated into their chromosomes, but the process has become clearer with further study. The bacteria become able to absorb DNA only toward the end of a period of vigorous growth. It appears that some apparatus for absorbing DNA is formed at the cell surface under the direction of certain genes. Once in the bacterium the DNA exchanges with part of the bacterial chromosome by recombination. It seems that ability to absorb transforming DNA can have survival value to the organism.

2. Bacteriophage Genes that Enter Infected Bacteria Are Made of DNA

In 1952, a few years after it became clear that DNA causes bacterial transformation, evidence was published that the genes of certain bacterial viruses, or bacteriophages, are made of DNA. The bacterial viruses that have been most studied are those that infect *E. coli* bacteria, and in particular the virus named T_2. The electron microscope shows that T_2 viruses have hexagonal bodies and a protruding tail. When a suspension of these viruses at 37°C is mixed with a suspension of *E. coli*, the viruses can be seen to become attached by their tails to the surfaces of the bacteria, where they remain. About 20 minutes later the bacteria burst and each releases about 100 complete new viruses, while the original infecting viruses can still be seen in outline attached to the bacterial membrane. It is clear that some compound (or compounds) passed from the infecting viruses into the bacteria where it caused the formation of the new viruses. This compound must, therefore, have contained the genes of the virus.

A first step to discovering of what compound these genes are composed can be made simply by analyzing the viruses. They are found to contain proteins and DNA but no detectable RNA; hence the genes cannot be made of RNA. Also, only the contents of the infecting viruses pass into the infected cells, and it is possible to release the contents artificially and analyze them. Viruses are suspended in concentrated salt solution, and water is then rapidly added. The viruses are burst by the change in osmotic pressure, and their contents are liberated, leaving the hexagonal coats, which can be separated by centrifuging (Figure 22). Analyses show that all the DNA of the viruses is liberated when they burst, while most of the protein remains with the coats. This suggests that the genes are made of DNA.

That the virus genes are made of DNA was proved almost conclusively by an experiment performed by A. D. Hershey and M. Chase at the Carnegie Institution in Cold Spring Harbor, New York. They prepared bacterial viruses in which the phosphorus and sulfur were radioactive. They did this by first growing *E. coli* bacteria in a nutrient solution containing radioactive phosphate and sulfate and then infecting the labeled bacteria with unlabeled viruses. A suspension of the radioactive viruses was then mixed with a suspension of ordinary *E. coli* cells. The viruses attached themselves to the bacteria, but after a few minutes they were broken away again by stirring the suspension in a Waring blender. Nevertheless, when a portion of the suspension was incubated for some minutes more, the bacteria burst, liberating the usual number of complete viruses. Hence some material must have passed from the viruses into the bacteria during the few minutes of attachment, and this material must have contained the virus genes. Hershey and Chase did not incubate most of the suspension after the viruses had been separated from the bacteria. Instead, they cooled and centrifuged it, so separating the bacteria from the remains of the viruses. They then measured the quantities of radioactive sulfur and phosphorus in each. They found that 85% of the radioactive phosphorus of the viruses had passed into the bacteria while 80% of the sulfur was still with the remains of the viruses. Now, the proteins of the virus contain sulfur but almost no phosphorus, while the DNA contains phosphorus but no sulfur, and there are no other important

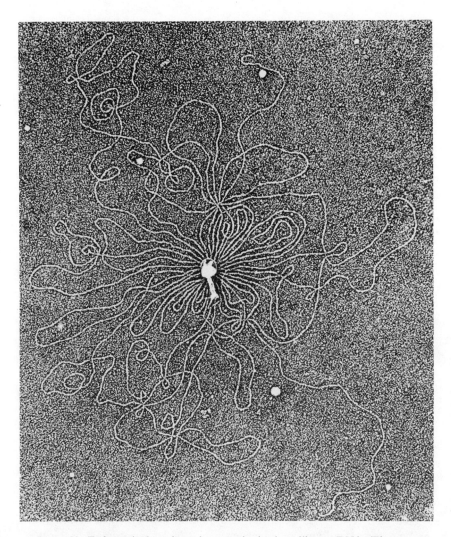

Figure 22. T$_2$ bacteriophage burst by osmotic shock to liberate DNA. (Electron micrograph courtesy of Drs. A. K. Kleinschmidt, D. Lang, D. Jacherts, and R. K. Zahn, Biochim. Biophys. Acta, 61:857, 1962, and Elsevier Publishing Co., Amsterdam.)

concentrations of phosphorus or sulfur in the virus. The bulk of the DNA must, therefore, have entered the bacteria while the bulk of the proteins did not. The experiments were later refined to show that almost all the DNA of the virus enters the bacterial cell on infection, but only 3% of the protein. There was therefore strong evidence that the virus genes are contained in the DNA that enters the bacterium, rather than in the trace of protein.

3. Some Virus Genes Are Made of RNA

A number of viruses contain no DNA but are composed of RNA and protein only. Examples are poliomyelitis virus, influenza virus, and the virus that causes mosaic disease in tobacco plants. Clear evidence has been obtained that the genes of these viruses are made of RNA.

A great advance in the understanding of viruses was made in 1935 when W. M. Stanley reported, from the Rockefeller Institute for Medical Research, "The isolation of a crystalline protein possessing the properties of tobacco-mosaic virus." When tiny quantities of these crystals were inoculated into leaves of tobacco plants, they became infected with mosaic disease. After a few days, crystals of the same "protein" could be isolated from the infected leaves in much larger amounts than had been inoculated. Stanley concluded, "Tobacco-mosaic virus is regarded as an autocatalytic protein which, for the present, may be assumed to require the presence of living cells for multiplication."

Many biochemists resisted Stanley's evidence that a virus could be no more than a pure chemical compound, as many had previously resisted evidence that enzymes are proteins. Their skepticism was not entirely unjustified, for in 1931, other workers had claimed the isolation of the same virus in the form of crystals, but these crystals had later been shown to be an inorganic compound that carried the virus as an impurity. As late as 1950, some reputable scientists still felt that Stanley's crystals were most probably a mere carrier of the true virus. However, in 1946, Stanley was awarded the Nobel Prize for his discovery, and the judgment of the awarders is now vindicated, for his crystals were undoubtedly pure tobacco-mosaic virus.

Work on the virus was continued by H. Fraenkel-Conrat and his colleagues at the Virus Laboratory in Berkeley, California, which was founded by Stanley. Analyses of Stanley's "protein" showed that it was not a simple protein but a ribonucleoprotein, that is, a non-covalent compound of protein and RNA. The electron microscope shows that the individual virus particles, from millions of which each of Stanley's crystals was built up, are tiny rods with protein on the outside and RNA threaded through the center. Fraenkel-Conrat's experiments were designed to discover whether the genes of this virus are made of RNA or protein or both.

He succeeded in separating the protein and RNA of the virus by mild procedures that left them undamaged. When these were mixed together under the correct conditions they, somewhat surprisingly, reformed the original viruses. This separation and reconstitution was done as follows (once again, years of hard work underlay a simple procedure). The protein was isolated by dissolving a sample of the viruses in alkaline buffer solution and leaving it for 2 to 3 days. The bonds between the protein and RNA were broken, and the protein could then be precipitated alone by adding ammonium sulfate solution of the correct strength. The RNA was obtained by another method. The virus was dissolved in a detergent solution and left at 40°C for 16 to 20 hours. The protein was then removed by adding ammonium sulfate solution, so leaving the RNA. Both the protein and RNA appeared not to be infective when inoculated into tobacco leaves. But if 1 ml of a 1% solution of the protein was mixed with 0.1 ml of a 1% solution of the nucleic acid, together with 0.01 ml of an acetate buffer, and the mixture was left at 3°C for 24 hours, the virus was reformed. This was proved by the ability to infect tobacco plants being very largely restored and by typical virus particles being visible in the electron microscope.

It might be expected from these experiments that the virus genes are made of a combination of RNA and protein, since neither compound alone appeared to be infective. But Fraenkel-Conrat proved in the following way that this is not so. The virus exists in a number of genetically different strains that can be distinguished by their differing effects on plants. He took two of these strains, which can also be distinguished by the protein of one containing certain amino acids that the other does not, and isolated protein and RNA from both. He then reconstituted hybrid viruses by combining the protein of one with the RNA of the other, and vice versa. Tobacco plants were then infected with these viruses and the progeny were examined. The result was beautifully clear-cut. Although hybrid viruses were used for infection, they were not produced in the plant. Instead, the viruses were of the strain from which the inoculated RNA had been isolated, and their protein contained amino acids characteristic of this strain. It was concluded that the genes of these viruses are made solely of RNA, that is, that the RNA controls the formation of all the inherited characters of the virus, including its protein of characteristic composition, and also controls the formation of more RNA of identical structure.

This conclusion was later confirmed by the discovery that pure RNA from tobacco-mosaic viruses will, in fact, give rise to normal infection if inoculated alone in sufficient quantity. The virus protein must, therefore, facilitate the infection of the plant without playing an essential role. Normal viruses are formed as progeny, complete with their characteristic protein. Recently, pure RNA has been isolated from a number of animal viruses and shown to be infective. For example, if RNA from polio virus is inoculated into human cells in tissue culture, the cells become infected and die. Normal polio virus containing both RNA and protein can be isolated from these cells. The genes of all these viruses must, therefore, be composed of RNA.

4. Evidence that Genes of Higher Organisms Are Made of DNA

Evidence for the importance of DNA in the chromosomes of higher organisms was obtained in the 1930s from studies of mutation, which were largely initiated by H. J. Muller at the University of Texas. When living organisms multiply, an individual occasionally appears that possesses a unit character in a form that was not present in its ancestors. This phenomenon is known as mutation and is found in all organisms, from viruses to man. An example of a mutation is the birth of lambs with short legs to sheep whose ancestors had legs of normal length, or the birth of a hornless calf to cattle whose ancestors all had horns. The mutant forms of these characters, once they have appeared, are continuously inherited, according to Mendel's principles. It is therefore clear that one or both of the pair of genes that control the character have undergone a permanent change; that is, a new allele has been formed. The mutant forms of the characters are usually recessive, and before they first appear the organism must inherit two genes of altered structure. Viruses and bacteria, in which only one gene controls each character, need inherit only one altered gene for a mutant form of a character to appear.

Mutations occur very seldom, and the rates at which they occur were at first difficult to measure. But Muller devised an ingenious system of breeding *Drosophila* by which the rate at which lethal mutations occur in one of its chromosomes could be measured. (A lethal mutation is one that kills an organism before its birth.) Using this technique, he made the discovery that the mutation rate was greatly increased if one of the flies was irradiated with X-rays before it mated. Many other plants and animals were soon irradiated by other workers and mutations were again found to be induced.

Ultraviolet light was also found to induce mutations, and it was a study of the effects of ultraviolet light that gave a clue to the importance of DNA in genes. Pollen grains of corn were irradiated under controlled conditions with ultraviolet light of different wavelengths. Each sample of pollen grains was used to fertilize corn plants, and the proportion of mutant plants in the progeny was measured. This proportion was found to depend directly on the extent to which the particular wavelength of light is absorbed by nucleic acids. Similar results were obtained with other organisms. These experiments did not prove conclusively that the genes of these organisms are made solely of DNA, since the light energy absorbed by the DNA might have been passed on to the true genes. But it was clear that DNA is at least closely associated with the structure of genes.

Many pieces of indirect evidence do, however, suggest that genes of higher organisms are composed solely of DNA. The chromosomes of higher organisms are composed of DNA, proteins (which are largely histones), and variable amounts of RNA. The possibility that genes contain RNA is made improbable by the fact that sperms contain insignificant amounts of RNA. The possibility that histones form an essential part of their structure is made unlikely by the fact that the proteins associated with DNA in sperm heads have a different composition from the histones of other tissues. DNA does not show these variations. It is always found in chromosomes, and elsewhere in the cell only in mitochondria and chloroplasts. Also, the quantity of DNA in almost every cell of a single species of animal or plant is the same, the most important exception being that germ cells contain exactly half this amount. This is the situation we should expect to find if each gene is made of a characteristic quantity of DNA and if germ cells contain half the number of genes of normal cells. Also, DNA from any cell of any member of a single species of animal or plant always contains the same relative amounts of the four component nucleotides. Differences in nucleotide composition begin to be detectable only as hereditary differences between organisms increase.

It is now quite certain that genes of higher organisms are composed solely of DNA. No definitive experiment, such as genetic transformation by DNA, has yet proved this. Our certainty comes rather from numerous facts that show that the chemical processes by which genes control unit characters are fundamentally the same in higher organisms as in bacteria and viruses. These facts will become evident in later chapters. This further similarity between higher and lower organisms has come as no surprise to biochemists, who already know how closely their metabolic reactions resemble each other.

5. Evidence that One Bacteriophage Gene Differs from Another in the Nucleotide Sequence of its DNA

We have seen that there is good evidence that the genes of most organisms are made solely of DNA. If we accept this as being so, in what way does one gene differ from another in structure? The differences in function of genes could conceivably be founded on one of two distinct kinds of chemical difference. First, each gene of an organism might be made from a number of DNA molecules, its characteristic function resulting from these molecules being arranged in a characteristic pattern. Alternatively, each gene might be a whole, or part of a whole, single DNA molecule of characteristic molecular structure on which its specific function depends. Important evidence that the second of these alternatives is correct came from experiments on recombination in bacterial viruses performed by S. Benzer at Purdue University around 1955.

These experiments were performed on the bacteriophage T_4, which is very like T_2, which was described earlier. When this virus is grown by incubating it with a suspension of E. coli cells in a nutritive medium, mutant viruses occasionally appear. A mutant that is found repeatedly is called the rII mutant. It differs from the normal virus in not being able to grow on a certain strain K of E. coli and in forming colonies of unusual appearance when grown on strain B of E. coli in agar jelly. Since each rII virus has a mutant genetic character, it must possess a gene that has a structure different from normal. A clue to what this difference in structure can be came from Benzer's very important discovery that there is not just one alteration in structure of the gene that gives rise to the rII characteristic but many. In other words, although rII mutants all look alike, the corresponding gene differs from the normal in these viruses in many different ways.

This could be demonstrated by recombination experiments in which E. coli bacteria were infected with, on the average, two rII virus particles per cell (see p. 24). If these two viruses were the progeny of a single rII parent, then the progeny of the double infection were (except for very occasional mutation back to the normal) all rII. But if the two viruses were progeny of rII viruses that had arisen by mutation on two different occasions, then normal viruses would usually be among the progeny. These could easily be counted by incubating the progeny with E. coli K cells on which only normal viruses will grow. It was evident that recombination had occurred between the chromosomes of the two strains of virus, part of the gene from one rII virus combining with part from another, to give a gene of normal structure. Hence, changes in structure that produce the rII characteristics must occur in various isolated regions of the gene, and chromosome breakage during recombination must occur within genes as well as between them. It is possible to make rII viruses arise more frequently than normal by treating the parents with certain mutagenic agents. In this way Benzer obtained over 300 rII viruses that were shown to differ from one another by recombination tests on E. coli. Hence the gene associated with this character can exist in over 300 distinct structures, which are precisely copied when the viruses multiply.

Benzer was also able to map the relative positions of these structural changes within the gene. He found that a large number of the 300-odd rII viruses could be arranged in a series that showed the following properties. On double infection in E. coli, virus 1 would recombine to give normal progeny with any others in the series except virus 2, virus 2 would do so with any except 1 and 3, virus 3 would do so with any except 2 and 4, and so on. It was concluded that these mutants each had a deletion (i.e., a piece missing) from a long or short segment of the length of the gene and that the deletions in successive mutants in the series overlapped. Hence their genes could not recombine to give an intact gene. An rII mutant that did not fall into this overlapping series would, on double infection of E. coli, give normal progeny with all but one, or a succession, of the members of this series, which therefore had a structural change in the same region. It was thus possible to make a map of the rII gene showing the position of structural change in each of the mutants. These mutations were arranged in one dimension, showing that the gene must be long and narrow.

We, therefore, have before us the following facts. This particular gene is, like others of the same virus, made of DNA. It is very long and narrow and can exist in over 300 different structural forms, each of which can be precisely copied when the viruses multiply. Also, it is possible to calculate roughly the quantity of DNA in the gene. The total DNA in the virus is about 100 million times the weight of a hydrogen atom. It follows from an estimate of the number of genes in the virus that the weight of each gene cannot be more than a few million on this scale and, hence, can contain only some thousand nucleotides. The almost inescapable conclusion is that the gene is either a whole, or part of a whole, DNA molecule, and that the variations in its structure that cause the appearance of the *r*II characteristic are changes in nucleotide sequence along long or short segments of the molecule.

Benzer's work was a thorough development of odd experiments that had already suggested that breakage can occur within genes during recombination, as well as between them, and that mutations can occur at different points along the length of a gene. For example, a gene controlling the red eye color of *Drosophila* will mutate to a recessive allele, for white eyes. In 1952, different strains of doubly recessive white-eyed *Drosophila* were crossed. When the resulting flies were interbred, a few normal red-eyed flies were found in the progeny. In these flies recombination had always occurred between genes on either side of the eye color gene, and hence chromosome breakage had occurred between them. It may be concluded that this breakage occurred within the eye color genes at such a position that an unmutated segment of one gene was recombined with a different unmutated segment of its partner. Similar but more comprehensive results were obtained with fungi at about the same time.

Therefore the reasonable working hypothesis can be drawn that every gene is either a whole, or part of a whole, DNA or RNA molecule and that one gene differs from the next in its nucleotide sequence. It will be seen that this hypothesis is fully supported by its ability to explain other facts of genetics in chemical terms.

6. Genes, Cistrons, and Complementation

Suppose that we had two strains of doubly recessive *Drosophila*, each showing a mutation in the same character. An example would be flies with "cherry" and "apricot" eyes instead of the normal dominant red. How could we determine whether both flies had a mutation in the same gene or in different genes that are both involved in controlling eye color? The standard way is by a *complementation test*. Flies that have the two mutant genes on different chromosomes are bred. If the genes are in fact alleles, these flies will display the character in a mutant form. Thus, flies with cherry and apricot on different chromosomes still have a mutant eye color. But if the mutations are on different genes, the flies will display the normal character, since each mutant gene will be *complemented* by a normal dominant gene on the homologous chromosome. Complementation tests can also be made in bacteria by introducing additional chromosome segments into the cell and in bacteriophages by introducing two phages into a single bacterium.

Using complementation tests, Benzer made an important discovery about the rII gene of T_4 bacteriophage. It will be remembered that an rII mutant will not grow on strain K of $E.$ $coli$ and for this reason recombination was brought about on strain B. But if strain K was infected with two different rII mutants, they did sometimes grow; that is, the mutants complemented one another. On the basis of complementation tests the rII gene could be divided into two sections: mutants with a structural change only in the first section could complement those with a change only in the second. Hence by traditional standards each of these sections would be considered to be a separate gene. But also by traditional standards they would be considered to be parts of the same gene since they lie adjacent on the chromosome with no clear break between them, and mutations within them produce the same change in visible character. Benzer therefore tried to introduce a term less ambiguous than $gene$. He called the two regions $cistrons$: two mutations are in the same cistron if there is no complementation when they are introduced into the same organism on separate chromosomes. The cistron is thus the basic functional unit of heredity since all its parts must be united if it is to function correctly. It will be seen later that the cistron does, in fact, produce a single chemical product. However, $cistron$, like most scientific terms, is not entirely without ambiguity, and $gene$ has tended to persist and usually refers to what Benzer meant by cistron.

Chapter 4

How Genes Make Copies
of Themselves

I. Watson and Crick's Structure of DNA

We have seen that Mendel's discoveries suggested an explanation of cellular heredity in chemical terms, namely, that genes can cause copies to be formed of their own component molecules and can also, directly or indirectly, control the formation, or entry into the cell, of all other molecules. With the knowledge that genes are made of DNA, the problem of their self-copying becomes clearer. Can DNA molecules in fact direct the formation of exact copies of themselves, and, if so, how? It is these questions that will be considered in this chapter.

Throughout the first half of this century the possibility was considered that self-copying, or autocatalytic, molecules exist, but precise ideas of what their structure might be, or how the self-copying reactions might occur, were lacking. Some scientists even built self-copying toys to assure their colleagues that the process was not impossible. Even by 1950, when it had become clear that DNA is probably the one important self-copying compound of living cells, there were no precise suggestions of the chemical reactions by which the process could occur. Then, in 1953, J. D. Watson and F. H. C. Crick, of Cambridge University, announced the most fundamental biochemical discovery of the century. It revealed that DNA molecules occur in nature in hydrogen-bonded pairs and that whatever the nucleotide sequence of one molecule of each pair may be, it always bears a set relation to that of its partner. The manner in which DNA molecules might form self-copies immediately became clear, and biologists were able to visualize the chemical process that underlies the division and self-copying of living cells that they had observed under the microscope for many years.

The manner in which Watson and Crick discovered this paired structure of DNA molecules is interesting because it concerns the correct balance between speculation and experiment in scientific research. Watson and Crick in fact performed no experiments on DNA. They made use of the published experiments of others, including those of M. H. F. Wilkins and his colleagues at London University, who were also trying to discover how DNA molecules are arranged in nature. But their suggested structure of DNA was largely the result of speculation. They knew that DNA molecules are chains of nucleotides, and they had certain clues from published experiments as to how these chains are arranged in relation to one another. From these facts, and much hard thinking, combined with the building of scale models of DNA molecules, they reached their conclusions about the probable structure of DNA. Their proposed structure was elegant and whether true or false threw great light on the type of chemical reactions by which the self-copying of DNA might occur. But when announced, it was not backed up by sufficient experimental evidence to make it conclusive. Within a few years, their structure was proved to be correct, largely by further careful X-ray diffraction measurements by Wilkins and his colleagues. They, in time, would no doubt have reached the same conclusions as Watson and Crick about the structure of DNA by a more orthodox approach, involving more experiment and less speculation. They could then have announced the discovery with strong experimental support.

Which approach to scientific discovery is the best? Obviously there is no straightforward answer. In the 1940s and 1950s biochemists were excessively cautious not to let speculation advance far beyond rigid experimental proof. This approach was largely a reaction against two ill-founded speculative bubbles of the 1930s, the cyclol theory of protein structure and the tetranucleotide theory of nucleic acid structure. However, lack of speculation can sometimes be a form of laziness. To many scientists, experiments are less painful to carry out than concentrated thinking. In recent years, under the influence of Crick, there has been a tendency for speculation among biochemists to increase, especially in molecular biology. So far, the results appear to be only good. The speculations have been carefully founded, and prophecies that experimental results would be tailored to fit them have not been justified. The Nobel awarders, at least, made no judgment as to whether speculation or experiment was more important in discovering the paired structure of DNA. In 1962 they awarded a prize for the discovery to Watson, Crick, and Wilkins.

When they started their speculations, Watson and Crick had a number of clues as to the way in which DNA molecules are arranged in relation to one another. First, many experiments had suggested that all samples of DNA are fibrous, that is, that the component molecules are stretched out side by side, and are not coiled into balls like the globular protein molecules discussed in an earlier chapter. Second, experiments had strongly suggested that hydrogen bonds occur in DNA and hence that they might bind one DNA molecule to another. Further experimental evidence also suggested that DNA molecules are arranged in solution so that the phosphate groups of each nucleotide lie exposed to the exterior, while the bases do not. Moreover, X-ray diffraction measurements hinted that, although DNA molecules are arranged lengthways, side by side, they are not stretched out as straight as they could be but are in the shape of long helixes, or corkscrews, and that each helix is made from two molecules.

From these clues, Watson and Crick tried to work out the precise way in which DNA molecules must be arranged in relation to one another. They realized that this arrangement is dictated by the sizes and positions of the constituent atoms. They therefore attempted to discover the arrangement by the most direct way: they built scale models of DNA molecules and tried to fit them together so as to satisfy the various clues as to DNA structure.

After some months of pondering and model building, Watson and Crick hit upon a most remarkable finding. They had discovered that two long DNA molecules could be fitted together side by side in the shape of a single right-handed helix rather like the strands of two-ply knitting wool, provided that they were arranged *antiparallel*: the 3' end of one molecule lying beside the 5' end of the other. Such a structure did have the phosphate groups of each nucleotide on the outside and the bases within. Most of the clues as to DNA structure were therefore satisfied. But experimental evidence had suggested that hydrogen bonds are present in DNA, and for the helix to be stable, the two molecules would need to be held together by hydrogen bonds. Watson and Crick found that hydrogen bonds could be formed between oxygen and nitrogen atoms of adjacent bases on the paired molecules but that these could be formed only between certain pairs of bases.

It will be remembered that the four nucleotides of DNA differ from one another in that their base may be adenine, guanine, cytosine, or thymine. (In some DNA, part of the cytosine has a methyl or hydroxymethyl group on its carbon 5. This does not affect any of the points that will be discussed here, and "cytosine" will mean cytosine and these derivatives.) It was found that a nucleotide containing adenine on one DNA molecule would, for reasons of space and orientation, only form hydrogen bonds with the adjacent nucleotide in the paired molecule if this nucleotide contained thymine. Similarly, guanine would form hydrogen bonds only with cytosine. Therefore, if the two molecules in the helix are held together throughout their length by hydrogen bonds between each adjacent pair of nucleotides, the nucleotide sequence of one molecule must bear a fixed relation to that of the other. Whatever the nucleotide sequence of one molecule, that of its partner would be given by the following simple rule: for A, substitute T; for T, substitute A; for G, substitute C; and for C, substitute G (where A = nucleotide containing adenine, and so on). Thus, if the sequence along part of one molecule happened to be –A–T–A–G–G–C–, that along the adjacent part of the paired molecule would be –⊥–∀–⊥–Ɔ–Ɔ–⅁–. (The second sequence is written upside down since it runs from the 3′ to the 5′ end because the molecules are antiparallel.)

Watson and Crick found that some strong support already existed for the idea that DNA always occurs with this precise relationship between the sequences of the paired molecules. Careful analyses had recently been made on many samples of DNA from different sources. These had shown that in any sample of DNA the number of residues of adenine and thymine were always equal, as were those of guanine and cytosine. Any deviations from this rule could be reasonably attributed to experimental error. Thus, DNA from salmon was found to contain adenine, thymine, cytosine, and guanine residues in the relative amounts 28 to 27 to 20 to 19. In DNA from a certain bacterium these ratios were 17 to 16 to 25 to 28.

To the biochemists who had made these analyses, the apparent equimolar ratios were something of a skeleton in the cupboard. About 10 years earlier, it had been claimed by certain biochemists that samples of DNA always contained residues of the four nucleotides in equal numbers. If the few analyses on which this claim was founded are examined, it is seen that there are quite large deviations from equality. But these deviations were thought to be due to experimental error, and on the basis of the analyses, the tetranucleotide hypothesis was fabricated. It suggested that DNA molecules always have the four nucleotides repeated one after the other, again and again, over their entire length. The painstaking analyses that exploded this hypothesis were the very ones that strongly suggested that adenine and thymine and guanine and cytosine residues occur in equal numbers. However, their discoverers were reluctant to proclaim them with any certainty lest they should fall into the same trap as their predecessors. Watson and Crick had no such inhibitions. They saw that the ratios emerged from careful analyses of many DNA samples, and they proclaimed them as strong support for their structure of DNA.

Details of the bonding between the paired molecules in Watson and Crick's structure of DNA are shown in Figure 23. (The helical structure itself is sometimes called the DNA molecule and is then said to be composed of two DNA *strands*.) It is important to emphasize again that the molecules in

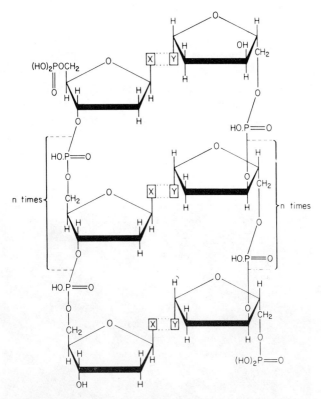

Figure 23. Two DNA molecules bound as suggested by Watson and Crick. X and Y represent bases linked by hydrogen bonds. When X is adenine, Y is thymine, and vice versa; when X is guanine, Y is cytosine, and vice versa.

each pair run in opposite directions, one end of one molecule being aligned with the opposite end of the other. Watson and Crick's structure allows the four nucleotides to be arranged in any jumbled order along one molecule and thus allows the existence of a very large number of kinds of DNA. But the sequence of the paired molecule must always be the translation of this jumble by the simple rule already mentioned.

It has now been proved beyond doubt that in nature DNA molecules are nearly always arranged in a structure almost identical to that proposed by Watson and Crick. (An exception is the DNA of the bacteriophage ϕX174, which consists of single unpaired molecules.) The strongest supporting evidence has been provided by comprehensive X-ray diffraction measurements made by Wilkins and his colleagues, which show that the distances between some of the atoms of the paired molecules are slightly different from those specified by Watson and Crick but that their structure is basically correct.

The structure of DNA varies slightly with humidity. That of the B form in which DNA occurs in nature is shown in Figure 24. Figure 25 shows details of the pairing by hydrogen bonds of adenine and thymine and of guanine and cytosine. This pairing occurs repeatedly in the chemical mechanism of genetics and is basic to it.

The pairing is reinforced by hydrophobic attractions between the

Figure 24. A model of a double helix. (By kind permission of Professor M. H. F. Wilkins.)

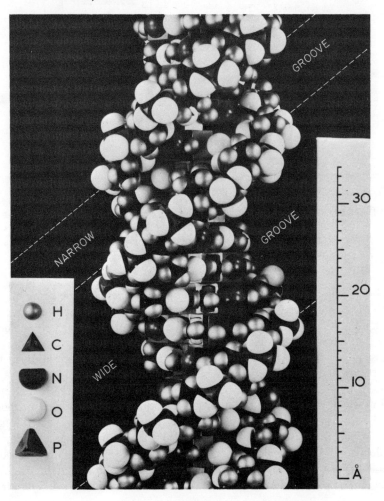

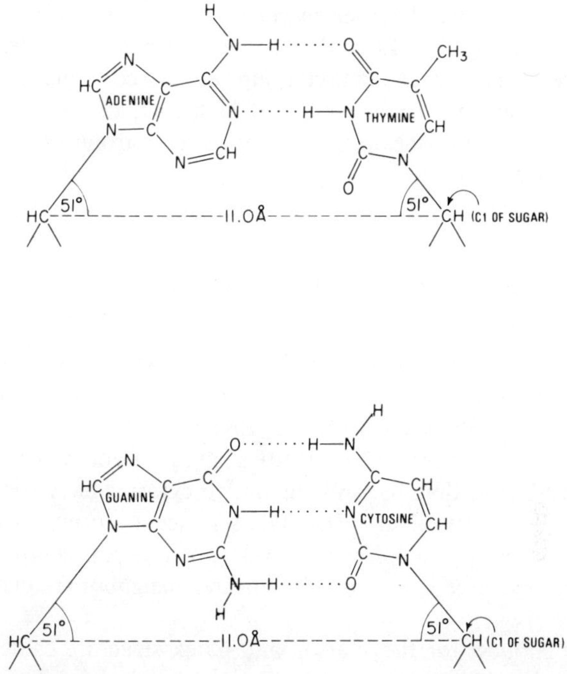

Figure 25. Adenine and thymine, and guanine and cytosine as hydrogen bonded in DNA.

bases, and consequently the structure of DNA is fairly stable to agents that weaken hydrogen bonds. It is seen that in each of the nucleotide pairs the distance between C–1 of the two sugars is the same (11.0 Å). Also, the angle between the bond linking C–1 of each sugar to its base and the line joining adjacent C–1s is the same (51°) for all four nucleotides. These facts mean that each sugar and phosphate residue is arranged in the same position relative to the axis of the helix no matter to which nucleotide it belongs. The helix therefore has the same dimensions throughout its length. Only consistent pairing between adenine and thymine and between guanine and cytosine can give this regular, and hence very stable, structure.

    The bases, the sugars, and the phosphate groups all recur at intervals of 3.4 Å along the helix axis. One turn of the helix contains 10 paired nucleotides and hence recurs every 34 Å (Figure 24). The diameter of the helix is about 20 Å. The bases lie with the planes of their rings roughly perpendicular to the helix axis, while the planes of the sugar rings are roughly in line with it. A narrow helical groove and a wide helical groove run the length of the structure. The narrow groove is the space between the paired molecules; the wide groove is the space between successive turns when the pair is wound into a helix (Figure 24). In DNA that contains 5-hydroxymethylcytosine (see p. 69), some of the hydroxymethyl groups are linked to glucose residues that lie in the wide groove. In chromatin of higher organisms it is believed that the basic proteins, histones and protamines, coil around the DNA helix in the wide or narrow groove.

2. Confirmation of Watson and Crick's Structure

This double-stranded structure of DNA is supported by numerous experiments. Comprehensive analyses show beyond doubt that samples of DNA almost always contain adenine and thymine and guanine and cytosine residues in equal numbers. Thus, in 101 analyses of different DNAs the mean ratios of adenine to thymine and guanine to cytosine have been found to be 1.009 and 1.001, respectively. That DNA molecules occur in pairs of equal length aligned in opposite directions with adenine paired with thymine and guanine with cytosine is confirmed by nearest-neighbor frequencies in DNA (see p. 78).

Early support for the Watson and Crick structure came from studies of the viscosity of DNA by C. A. Thomas at Harvard University. DNA gives very viscous solutions in water, as do other compounds whose molecules are long and narrow and tend to entangle with one another. Normally, if the molecules in a solution of such a compound are gradually made shorter, the viscosity of the solution falls proportionally. DNA molecules can be shortened by incubating them with an enzyme deoxyribonuclease, which catalyzes the breaking of the linkage between two successive nucleotides. It acts more or less at random along the length of the molecule and shows little preference for the long DNA molecules or for the fragments formed by their breakdown. However, when a solution of DNA is incubated with this enzyme, the viscosity does not fall directly with the number of bonds between nucleotides that are severed. Instead, it remains high for a much longer time than would be expected and then falls rapidly.

This delayed fall in viscosity can be perfectly explained by the pairing of DNA molecules as suggested by Watson and Crick. To understand why, it is helpful to compare each pair of DNA molecules to a ladder in which the uprights on either side represent the two DNA molecules and the rungs represent the hydrogen bonds between adjacent nucleotides. (Unlike ladders, DNA molecules are, of course, coiled into helixes, but this does not affect the argument.) Suppose that both uprights of a ladder are sawed through separately at random somewhere along their length. The ladder will only be shortened in length if both cuts have been made between a single pair of rungs, and the chance of this happening is small. If a series of similar cuts in each upright is made, the chance of a cut occurring opposite one that has already been made, and hence of the ladder being shortened, will increase continuously. The delayed fall in viscosity when solutions of DNA are incubated with the enzyme can be explained by a similar principle. The chance of any two paired molecules being severed between a single pair of hydrogen bonds is small at first, but as incubation proceeds the chance of a breakage occurring opposite one that already exists will increase. Thomas showed that the rate at which the viscosity of DNA solutions falls can be exactly explained in this way, with one reservation. A pair of DNA molecules appears to be shortened in length not only when both are severed between the same pair of hydrogen bonds but also when the breaks are on either side of a common hydrogen bond. This may be compared to a ladder becoming shortened when the uprights are severed on either side of a common rung that is then torn away.

Other early evidence in support of Watson and Crick's structure of DNA came from experiments by P. Doty and his colleagues at Harvard University. They found that if samples of DNA are heated in dilute salt solutions for a few minutes, their properties change in a way that can be explained only by paired DNA molecules becoming wholly, or partly, separated. For example, their appearance in the electron microscope changes. The long, straight, narrow strands become shorter and more irregular as would be expected if the DNA molecules were no longer supported by mutual hydrogen bonding. The change is known as *denaturation.*

At first sight it appears surprising that the paired DNA molecules can be so easily separated from the helical structure, but on considering the forces involved this is not so. That the separated molecules will just as easily recombine to give the perfect double helixes seems at first sight ever more improbable, but on further thought it is clear that the process is not unlike the formation of crystals. Doty realized this and chose his experimental conditions accordingly. He warmed solutions of the separated molecules to temperatures just below those needed to separate them originally, and the molecules recombined into double helixes. Evidence of this renaturation came partly from the molecules regaining their original appearance in the electron microscope, but other elegant proofs were devised. In one experiment, for example, it was found that DNA from a certain strain of bacteria, which was able to cause genetic transformation of a related strain, lost this ability when the paired molecules were separated from one another by heating. When the solution of the separated molecules was warmed to a temperature that was believed to cause them to recombine, the transforming ability was largely restored.

The temperature at which this denaturation occurs under standard conditions is now used to characterize a sample of DNA. The denaturation of a solution of DNA is accompanied by a rise of about 40 % in its absorption of ultraviolet light, known as the *hyperchromic effect*. The middle of the temperature range over which the rise occurs is known as the *melting temperature* (T_m), and in solutions of defined pH and salt concentration, it is characteristic of each DNA. The greater the content of guanine plus cytosine (which can, in fact, be determined from the T_m), the higher is the T_m. The greater stability of the guanine-cytosine pair has been attributed to its having three hydrogen bonds rather than two (Figure 25), but it appears that hydrophobic attractions are also greater between guanine and cytosine than between adenine and thymine. Renaturation of denatured DNA is usually done by holding the DNA in a solution of high ionic strength at about 25°C below T_m. If this renaturation is carried out in the presence of RNA of the same base sequence as one or both strands of the DNA, DNA:RNA hybrid double helixes are formed. As will be seen later (p. 102) this *hybridization* is a valuable test of gene function.

3. The Self-copying of DNA In Vitro

The importance of Watson and Crick's proposed structure was that the relation between the nucleotide sequences of the paired molecules immediately suggested how DNA might direct the formation, within living cells, of precise copies of itself. Suppose that the paired molecules of a DNA double helix became separated from one another, by the breaking of the hydrogen bonds that joined them, and that the region of the cell in which this occurred contained each of the four nucleotides, or molecules closely related to them, in solution. These free nucleotides would tend to become attached by hydrogen bonds to the nucleotides of the two separated DNA molecules. Free nucleotides that contained adenine would become bound to those in DNA that contained thymine, and so on. If an enzyme was present that would catalyze the linking together of the free nucleotides when hydrogen-bonded to a single DNA molecule, a new DNA molecule of the correct sequence would be formed beside the first. The replication of DNA would occur by each molecule directing the formation of a copy not of itself but of the partner from which it had separated. Every DNA molecule would thus depend on a symbiotic relationship with its partner for its nucleotide sequence to be carried on into subsequent generations of cells.

Does DNA, in fact, form copies, essentially by this mechanism, which was proposed by Watson and Crick? Evidence that it does was first provided by A. Kornberg and his associates, and it will be discussed in this section. Their experiments were performed largely at Washington University, St. Louis, and at Stanford University, and for them Kornberg was awarded the Nobel Prize in 1959. The experiments were designed to isolate an enzyme from living cells that would catalyze the self-copying of DNA *in vitro*. Kornberg chose cells that he considered most likely to contain this enzyme in high concentration: *E. coli* that were dividing every 20 minutes and hence rapidly forming DNA. He disintegrated some of these bacteria in a buffer solution, hoping in this way to release the enzyme or enzymes involved in DNA synthesis. He then added the compounds that he thought would most probably be used by the cell to make DNA. These were not the component nucleotides themselves, but rather compounds with two additional phosphate groups attached to the phosphate of the nucleotide, that is, the nucleoside triphosphates. From a comparison with similar reactions in living cells, it was expected that these additional phosphate groups would "activate" the nucleotides, which could then combine, under the influence of the enzyme, to give a high yield of DNA, eliminating the additional phosphates as pyrophosphate in the process.

Kornberg realized that even if these compounds were converted to DNA by the disintegrated bacteria, no increase in DNA would be measurable, because the DNA of the bacteria was being rapidly degraded by other enzymes. He therefore incorporated radioactive carbon 14 into one of the nucleotides, and after incubation he isolated the DNA of the disintegrated bacteria to see if it had gained any radioactivity, and hence if any synthesis had occurred against the tide of degradation. He found only traces of radioactivity, but by subdividing the proteins of the bacteria into a number of fractions, he obtained a fraction that would catalyze the formation of DNA of high radioactivity. This contained the synthesizing enzyme largely separated from the enzymes that caused DNA degradation. He then set about studying the properties of this purified enzyme, which he named DNA polymerase.

He found that it had a property that marked it off from all other enzymes. It would only catalyze the rapid synthesis of DNA when incubated with the triphosphates of all four nucleosides if a small quantity of DNA was first added as *template*. The template DNA (which consisted of paired molecules of complementary nucleotide sequence) did not need to come from *E. coli*. Kornberg produced overwhelming evidence that the DNA that was formed, and that could be over 20 times the quantity of the template DNA, always consisted of paired molecules of the same nucleotide sequence as the template DNA. In other words, the enzyme had the unique property of catalyzing the self-copying of the template DNA.

Kornberg's evidence that the structures of template and product were always identical was of many kinds. The relative numbers of the four nucleotides in each were shown to be identical in a number of experiments. For example, when a bacterial DNA, used as a template, contained adenine, thymine, guanine, and cytosine nucleotides in the ratio 0.65 to 0.66 to 1.35 to 1.34, the product contained them in the ratio 0.66 to 0.65 to 1.34 to 1.37.

When the template was DNA from cattle, in which these ratios were 1.14 to 1.05 to 0.90 to 0.85, those in the product were 1.12 to 1.08 to 0.85 to 0.85. Any differences here between the ratios in template and product could easily be due to experimental error (as, incidentally, could any deviations from a 1 to 1 ratio between adenine and thymine and between guanine and cytosine nucleotides). Kornberg still found these identical ratios in template and product even when the molecular ratios between the nucleotide phosphates on which the enzyme acted were varied widely. If one of the phosphates was withheld, no DNA self-copying occurred. It was also shown that the DNA formed by the enzyme had the same viscosity as the template and that the viscosity showed the same delayed fall as that of the template on incubation with deoxyribonuclease.

Kornberg also devised a method of *nearest-neighbor frequency analysis* that proved that the product of the DNA polymerase reaction contained paired molecules aligned in opposite directions with base pairing between adenine and thymine and between guanine and cytosine. In this method one of the nucleoside triphosphates used in the enzymic synthesis of DNA is labeled with ^{32}P in the α-phosphoric acid residue. This residue is linked directly to carbon 5 of the sugar and is not eliminated as pyrophosphate in the reaction but forms a second ester link with carbon 3' of the neighboring nucleotide in the DNA that is synthesized (Figure 26). The labeled DNA formed in the reaction is isolated and enzymically hydrolyzed to the 3'-phosphates of the component nucleosides. The original bond from the ^{32}P to carbon 5' of the parent nucleoside is thus severed, and it is left attached to carbon 3' of the nearest neighbors. The ^{32}P contents of each of the four nucleoside 3'-phosphates are measured, and these are proportional to the frequencies with which they occurred next to the nucleotide that was labeled. The procedure is repeated with each of the four nucleotides in turn labeled with ^{32}P. In one experiment Kornberg carried out an analysis of nearest-neighbor frequencies using DNA from *Mycobacterium phlei* as template. Since there are 4 nucleotides in the DNA, there are 16 possible dinucleotide sequences or *nearest-neighbor pairs*, namely, A–A, A–T, A–G, A–C, T–A, T–T, T–G, T–C, G–A, G–T, G–G, G–C, C–A, C–T, C–G, and C–C. (When we write A–T, for example, this means that the nucleotide whose base has the initial letter A is linked from carbon 3 of its sugar through phosphate to carbon 5 of sugar of the nucleotide whose base has the initial letter T. See also caption to Figure 18, p. 45.) With *M. phlei* DNA as template, Kornberg found the following values for nearest-neighbor frequencies: A–A, 2.4%; T–T, 2.5%; C–A, 6.2%; T–G, 6.3%; G–A, 6.6%; T–C, 6.1%; C–T, 4.5%; A–G, 4.5%; G–T, 6.0%; A–C, 6.4%; G–G, 8.9%; and C–C, 9.0%. The values shown are the percentages that each nearest-neighbor pair forms of all the pairs that occur along the molecules of newly synthesized DNA. (In the molecule G–C–C–A–T–A, for example, all the pairs that occur along

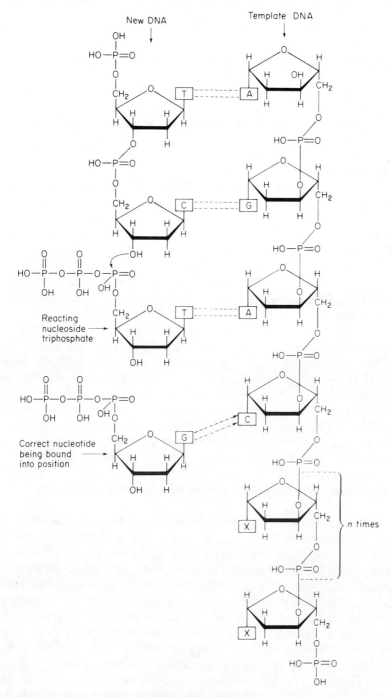

Figure 26. The synthesis of DNA by Kornberg's polymerase. *A, G, C,* and *T* represent the bases adenine, guanine, cytosine, and thymine. (*X* represents any one of these four bases.)

the molecule are G–C, C–C, C–A, A–T, and T–A.) It is seen that, within the limits of experimental error, successive frequencies in the above list are equal to one another. This result proves that the DNA synthesized by DNA polymerase, which has the same composition as the *M. phlei* DNA template, consists of paired molecules of equal length aligned in opposite directions with adenine paired with thymine and guanine with cytosine.

Kornberg's experiments therefore proved conclusively that DNA can in fact direct the formation of self-copies (although it is now clear that the DNA synthesized *in vitro* does not become separated from the template as discrete molecules of identical length but branches out from it). When Kornberg's DNA polymerase was first discovered, it seemed certain that this enzyme was involved in the self-copying of all genes that precedes cell division. However, in 1967, J. Cairns, of the Cold Spring Harbor Laboratory in New York, provided strong evidence that it is not. He isolated a mutant strain of *E. coli* that could grow and hence replicate its genes, but from which none of Kornberg's DNA polymerase could be isolated by the normal procedure or by any other. This strain of *E. coli* was particularly susceptible to mutation by ultraviolet light, and it appears that Kornberg's DNA polymerase is responsible for repairing DNA that has been altered in chemical structure by radiation. If part of one of the paired strands of DNA is altered, this can be excised enzymically and a new section built up against the complementary strand by Kornberg's polymerase, the ends then being joined to the original strand by another enzyme. The enzymes responsible for DNA replication have still not been fully identified, although the chemical reactions that they catalyze are certainly closely related to those in DNA repair.

4. The Strands of a DNA Double Helix Pass into Different Copies

Kornberg's experiments proved conclusively that DNA can direct the formation of self-copies, but they gave no evidence as to the truth of Watson and Crick's suggestion that during self-copying the paired molecules or strands of the DNA double helix separate from one another, each forming one half of a new double helix. Evidence that they do was first provided by an experiment performed in 1958 by M. Meselson and F. W. Stahl at the California Institute of Technology. To understand their experiment, it is necessary to consider the destiny of the two strands of any one DNA double helix when continued replication occurs according to Watson and Crick's mechanism. The two strands separate and a new complementary strand is formed beside each of them, so giving two double helixes identical to the first. Each of these helixes contains one of the original strands and one new one. Suppose that these two helixes now give four single strands from which four new double helixes are formed. Two of these helixes still contain one of the original strands, while the other two do not. Forever after, if these 4 double helixes give 8, and the 8 give 16, and so on, two of the helixes will always contain one of the original strands.

These facts are a consequence of self-copying of DNA in the manner proposed by Watson and Crick. Even if their double helix is accepted as the correct structure of DNA, it is conceivable that self-copying might not occur in this way. For instance, each strand of the double helix might direct the formation of a new complementary strand but then return to pair with its original partner.

Meselson and Stahl devised an experiment to follow the destiny during repeated replication in *E. coli* of DNA molecules that had been iso-topically labeled and to discover whether it accorded with the predictions of Watson and Crick. They obtained bacteria that contained labeled DNA by growing them in a culture solution containing glucose, mineral salts without nitrogen, and ammonium chloride in which almost all the nitrogen atoms were the heavy isotope ^{15}N. They incubated the culture until it contained 14 times as many bacteria as had been inoculated. Hence virtually all the nitrogen ($^{13}/_{14}$) in these bacteria was heavy nitrogen. A portion of the bacteria was separated by centrifugation, and their DNA was isolated. To the remainder, a culture solution containing normal ammonium chloride was added in large excess, and the incubation was continued. The bacteria whose DNA contained only heavy nitrogen continued to multiply and form new DNA from ammonium chloride that contained only normal nitrogen. Samples of the bacteria were withdrawn when they had doubled and further increased in numbers, and DNA was isolated from each.

Meselson and Stahl devised an ingenious way of comparing the densities of the paired DNA molecules in these samples and their relative contents of heavy nitrogen. If a cesium chloride solution is spun in an ultracentrifuge at a speed that gives 140,000 times the force of gravity, its molecules begin to sediment. However, because their molecular weight is relatively low, they never separate completely. Instead, after 20 hours, a steady state is reached in which the concentration of cesium chloride, and hence the density of the solution, gradually increases toward the bottom of the tube. If DNA is added to the solution before centrifuging, it comes to rest in a band in a region of the tube where its density equals that of the solution. The position of this band can be found by photographing the tube in ultraviolet light. The content of DNA can be found from the extent of the light absorption. It was found that the DNA containing only heavy nitrogen formed a band lower down the tube than did that containing only normal nitrogen.

In this way, they centrifuged their samples of DNA, which had been isolated at various times after the bacteria containing heavy nitrogen had continued their growth in the solution containing normal nitrogen. It will be remembered that one of these samples was isolated after the bacteria had doubled their numbers, and hence doubled their quantity of DNA, in this solution. It was found that this DNA contained particles of only one kind, which were intermediate in density between those from DNA containing only heavy nitrogen and only normal nitrogen. This finding is in accord with Watson and Crick's method of replication. This DNA should be composed of hydrogen-bonded pairs of one heavy and one light strand, and Meselson and Stahl proved that it was by heating the DNA to about 100°C and centrifuging the separated molecules in the cesium chloride density gradient: the DNA now consisted of equal numbers of heavy and light particles. The DNA isolated after further growth also contained particles of the correct densities in precisely the relative numbers predicted. For example, when the bacteria had quadrupled in number, their DNA contained particles of two densities. One half had the intermediate density, while the remainder had the density of DNA containing only normal nitrogen. It is clear that these experiments exclude certain conceivable methods of DNA replication and are fully in accord with the mechanism of Watson and Crick. These experiments were later repeated on the DNA of human cells in tissue culture, and of *Chlamydomonas* and certain bacterial viruses, with the same results.

Around 1957, J. H. Taylor of Columbia University obtained similar evidence that the DNA of higher organisms replicates according to the mechanism of Watson and Crick. In higher organisms the DNA of each chromosome doubles in quantity in the S phase, a period between cell divisions when the chromosomes are invisible. When they reappear before mitosis, each chromosome appears to have divided into two daughter chromosomes or *chromatids*. Taylor's experiments were designed to discover how the newly formed DNA is distributed between the two daughter chromosomes, and between the chromosomes formed from each of them by subsequent replication. He performed his experiments on root tips of plants in which the cells were continually dividing. He immersed these root tips for a short while in solutions of the nucleoside thymidine, which was labeled with radioactive hydrogen, ^{3}H. This is converted into one of the four nucleoside triphosphates and incorporated into DNA by plant cells. Taylor left the root tips in this solution long enough for the DNA of many cells to double in quantity but not long enough for it to double yet again. Any DNA that was formed during this doubling was radioactive. But after removing the root tips from the radioactive solution, they were washed so that any DNA formed subsequently would not be radioactive.

The root tips were then left for a few hours to allow some of the pairs of daughter chromosomes, which had been formed in the radioactive solution, to become visible. Did both daughter chromosomes of each pair contain radioactive DNA, or was it confined to one of them? Taylor answered this question by the technique of autoradiography. He flattened the cells on microscope slides and, in the dark, placed photographic film against them. After a while he developed the film. Wherever a chromosome had lain against the film its outline could be seen under the microscope. If this chromosome contained radioactive DNA, its outline was scattered with black dots of silver grains where radiations had fogged the film. Taylor found that both members of each pair of daughter chromosomes contained radioactive DNA, and counts of the black grains showed that they contained it in equal amounts. This is the result predicted by the mechanism of Watson and Crick and not by certain other self-copying mechanisms. On Watson and Crick's theory, when every DNA double helix in the parent chromosome replicates, it gives two helixes that contain one old and one new strand each—in this experiment, one inactive and one radioactive strand. When the double helixes become evenly distributed between the daughter chromosomes, these will acquire equal amounts of radioactivity. If each chromosome contains only one large double helix, Watson and Crick's theory predicts a quite different distribution of radioactivity if these radioactive daughter chromosomes are themselves left to duplicate in a solution that is not radioactive. In the radioactive daughter chromosomes one strand of the double helix will be radioactive and the other inactive. When this DNA replicates, it will give two helixes, only one of which contains one radioactive strand. Hence, when this chromosome divides, only one of its daughters will receive a radioactive molecule.

Taylor tried to discover whether this uneven distribution of radioactivity among the progeny of the radioactive daughter chromosomes does, in fact, occur. He allowed cells that contained radioactive daughter chromosomes to go through another complete cycle of chromosome duplication in a solution that contained no radioactivity. He then carried out autoradiography when the duplicated chromosomes reappeared. He found that almost all pairs that were formed from a radioactive parent contained one active and one inactive chromosome, in excellent agreement with the predictions of Watson and Crick. Similar experiments have since been made on the chromosomes of human cells in tissue culture with the same results.

At the time of these experiments it seemed highly improbable that a chromosome could contain only one DNA double helix, since this would be some hundred times the length of the chromosome. It was thought that the DNA was subdivided into smaller double helixes, and to explain the ordered segregation of radioactivity, it was necessary to assume that these were linked in some ordered way to a protein backbone in the chromosome. The results are much more readily interpreted if, as now seems probable, there is only one double helix in each chromosome.

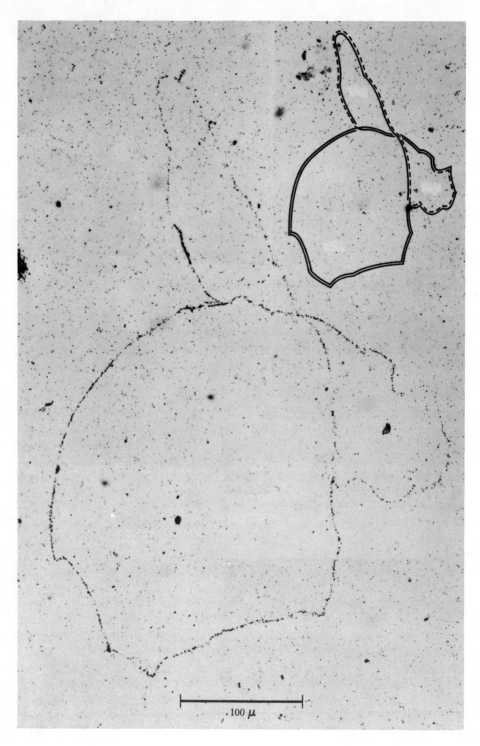

100 μ

5. New DNA Strands Are Formed as the Old Ones Separate

The experiments of Meselson and Stahl and others prove that when DNA self-replicates the two strands of the double helix pass into different daughter molecules. Do the strands separate completely before any replication occurs, or does replication occur along the single strands as these are drawn apart from the double helix? In 1962, J. Cairns prepared pictures of chromosomes of *E. coli* in the process of replication that show clearly that DNA synthesis occurs along both strands of the double helix as they separate from one another at the *replicating fork*. He obtained these pictures not by direct electron microscopy but by autoradiography: the bacteria were grown in a medium with (^{3}H)thymidine, and their labeled chromosomes were isolated and covered with photographic emulsion on a microscope slide. On developing the slide, the structure of the chromosomes was revealed under the light microscope by narrow lines of silver grains. Since the chromosome of *E. coli* is about 1 mm long and only 20 Å or 2×10^{-6} mm wide, it was very easily broken during isolation. Cairns minimized the breakage by mixing the bacteria with unlabeled DNA and then rupturing them with the enzyme lysozyme and a detergent. The bacterial chromosomes were released and supported by the excess of unlabeled DNA, which was collected on a membrane.

Cairns' pictures showed that the *E. coli* chromosome is a single DNA double helix and is circular, as had been suggested by mapping experiments (p. 23). In bacterial viruses the circle is closed by the same internucleotide links as in the remainder of the DNA, so that the double helix is endless, and the DNA of *E. coli* is probably similar. The autoradiographs show that DNA replication occurs at one site, which travels around the DNA circle as a replicating fork, separating the two strands of the double helix, on each of which a new strand is immediately formed (Figure 27). Two important problems are raised by these facts. First, because the two strands of a DNA double helix are aligned in opposite directions (see p. 69), different reactions are needed to extend each of the new strands in the direction of the replicating fork: only the new strand that has a free 3'-hydroxyl at the replicating fork can be extended in that direction by an enzyme that catalyzes the reactions of

---

Figure 27. Autoradiograph of replicating *E. coli* chromosome that was derived from a parent chromosome that replicated in the presence of (^{3}H)thymidine; hence it had one labeled and one unlabeled strand. It was also itself in the middle of replication in the presence of (^{3}H)thymidine. Grain counts show that at the replicating fork one new double helix has two labeled strands and that the other one has one, as expected. (Courtesy of Dr. J. Cairns.)

Kornberg's polymerase (see Figure 26, p. 79). The second problem is that of unwinding of the double helix as replication occurs. If the strands of two-ply knitting wool are pulled apart from one end, the other end rotates continuously. But if the ends of each strand were joined into a circle, this essential rotation would be prevented. Hence the circular DNA double helix must in some way be opened to allow rotation as replication occurs. How these two problems are, in fact, overcome is not yet understood.

Chapter 5

How Genes Control the
Formation of Other
Cell Molecules

I. One Gene to One Protein

It has been seen that DNA doubles in quantity between cell divisions by forming self-copies. It will be seen in this chapter that other molecules of cells are not newly formed in this way but that their formation is either directly or indirectly controlled by the DNA of the genes. This conclusion arose largely from the work of G. W. Beadle and E. L. Tatum, at the California Institute of Technology, for which they were awarded the Nobel Prize in 1958. Their experiments were designed to discover how each gene exerts its effect on a genetic character and suggested that the primary action of each gene is to control the formation of a protein of specific structure.

The first clue to this action of genes came around 1909 from investigations by Sir A. E. Garrod in England on "inborn errors of metabolism," namely, diseases that certain people inherit in which certain chemical reactions of the body are abnormal. An example is alcaptonuria, in which the urine turns black in air because it contains an abnormal compound, homogentisic acid. Garrod gave evidence, from the occurrence of the disease in families, that it is the recessive form of a single unit character. He suggested that homogentisic acid is always produced in the body but is normally converted to some other compound. Thus the dominant form of the character would be the ability to bring about this conversion. Garrod made the important suggestion that in this and similar diseases "the most probable cause is the congenital lack of some particular enzyme, in the absence of which a step is missed, and some normal metabolic change fails to be brought about." However, he did not formulate a general hypothesis that a dominant gene acts by controlling the formation of a particular enzyme or that a recessive gene is faulty in this respect. It is probable that he considered this action limited in extent.

In 1935 Beadle began experiments directly designed to discover the chemical reactions by which genes control inherited characters. He first studied eye color in *Drosophila* and obtained evidence that each gene involved controls one chemical reaction by which eye colors are formed. However, he made more rapid progress a few years later when he collaborated with Tatum. The difficulty of studying most inherited characters in this way is that although they must result from the formation of chemical compounds these are unknown and must be discovered before the primary chemical action of the genes can be investigated. Tatum and Beadle decided to circumvent this difficulty by studying inherited characters that are themselves clear-cut chemical reactions, namely, the ability to make certain vitamins and amino acids from simpler compounds. For their experiments they chose the mold *Neurospora*, which has many valuable characteristics for genetic experiments.

A culture of *Neurospora* consists of a mass of ramifying hyphae that contain haploid nuclei with seven unpaired chromosomes. From these hyphae are budded off spores that contain one or more haploid nuclei and that will germinate and divide mitotically to give more hyphae. However, sexual reproduction also occurs. Female organs branch off the hyphae as filaments, and the haploid nuclei from spores will fuse with them, provided that the spores are from another mold that differs in a single gene that determines the mating type. A haploid nucleus formed by mitotic division of the male spore nucleus then fuses with one derived from the female organ. This diploid nucleus immediately undergoes meiosis. As usual (see p. 13), in the first division of meiosis, homologous chromosomes from the paternal and maternal nucleus come to lie side by side in pairs (each chromosome having already divided into two chromatids). The chromosomes of each pair then separate into two haploid nuclei, which each contain a random assortment of paternal and maternal chromosomes. These nuclei are placed at either end of a narrow cell, and after the chromatids of each chromosome separate at the second division of meiosis the four resulting nuclei lie in a row. These four nuclei then divide once more to give a row of eight nuclei, which become a row of eight spores (Figure 28).

The fact that the origin of these eight spores is known is one of the

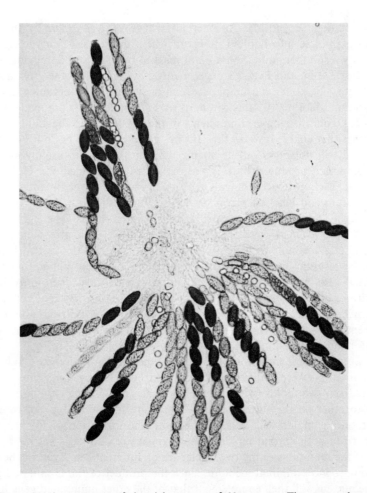

Figure 28. Arrangement of the eight spores of *Neurospora*. The spores shown result from crossing a normal mold with one that produces unpigmented spores. Separation of pairs of dark spores from one another results from crossing over. (Microphotograph courtesy of Dr. D. R. Stadler.)

valuable features of *Neurospora* for genetic studies. Each spore can be removed with a dissecting microscope and grown into a mold whose inherited characters can be identified. Among other things, this allows conclusions to be drawn about chromosome behavior in meiosis. For example, recombination between a pair of paternal and a pair of maternal alleles is usually only observed in molds grown from two of the four pairs of spores. Hence, each crossover must involve only one of the two chromatids of each of the pairing chromosomes (see p. 14). Another advantage of *Neurospora* is that the spores germinate into haploid molds in which the recessive characters that normally result from mutation are not masked by unaltered dominant genes.

For Tatum and Beadle's experiments *Neurospora* had the great advantage that it will grow on a simple medium of sucrose and inorganic salts plus one vitamin, biotin. All other vitamins and all amino acids are made within the mold from the simple components of the medium, and the mold inherits the genes that enable it to do this. If one of these genes is put out of action, the mold will not grow on the simple medium, but it will usually do so if the compound produced by this gene is added to the medium. Hence the chemical reaction that is controlled by this gene can be readily identified.

Tatum and Beadle irradiated spores of the mold with X-rays or ultraviolet light and used them to fertilize unirradiated molds. The eight spores formed after meiosis were then dissected out and each grown in a tube containing a medium with many amino acids and vitamins added. Spores from the resulting molds were then tested for growth on minimal medium. If they did not grow, amino acids and vitamins were then added individually to cultures of more spores to discover which compound they required. After culturing around 70,000 spores, Tatum and Beadle isolated about 400 different mutant strains of the mold that required addition of a growth factor to the minimal medium.

Some of these mutant molds required the same growth factor. Whether their mutations were in different genes could be tested by complementation. Spores from two molds (which must be of the same mating type) are placed close together on the surface of a solid culture medium. Hyphae from the two resulting molds will usually fuse, with the formation of hyphae with nuclei from both mutant molds. These hyphae are known as *heterokaryons*. Only if the mutations in the two nuclei are on different genes will the heterokaryon grow on minimal medium, since the product of the unmutated gene in one nucleus will then complement that of the other unmutated gene in the other nucleus. The relative positions of these mutations can also be mapped by recombination measurements in which two molds are crossed and the frequency is measured at which spores are formed that will grow on minimal medium.

Many of the mutant molds that required the same growth factor did in fact have mutations in different genes. Hence it was clear that a number of genes are concerned in the synthesis of any one of these compounds. Tatum and Beadle, and many other workers, concluded that these genes each control the formation of a distinct enzyme. The following is the type of evidence on which this conclusion was based. Three strains of *Neurospora* were isolated that required the amino acid arginine and hence could not make it from the components of the minimal medium. Each strain had a mutation in a different gene. One of the strains would grow if arginine or the amino acids citrulline or ornithine was added. Another would grow with arginine or citrulline but not with ornithine, while the third would grow only with arginine. It was known that in liver arginine is formed from citrulline, which is formed from ornithine, and it was proved that the same reactions occur in *Neurospora:*

$$
\begin{array}{ccccc}
 & & NH_2 & & NH_2 \\
 & & | & & | \\
 & & C{=}O & & C{=}NH \\
 & & | & & | \\
NH_2 & & NH & & NH \\
| & & | & & | \\
CH_2 & & CH_2 & & CH_2 \\
| & & | & & | \\
CH_2 & \longrightarrow & CH_2 & \longrightarrow & CH_2 \\
| & & | & & | \\
CH_2 & & CH_2 & & CH_2 \\
| & & | & & | \\
CH.NH_2 & & CH.NH_2 & & CH.NH_2 \\
| & & | & & | \\
COOH & & COOH & & COOH \\
\\
\text{Ornithine} & & \text{Citrulline} & & \text{Arginine}
\end{array}
$$

Hence it was clear that the first strain lacked the ability to make ornithine, the second to convert ornithine to citrulline, and the third to convert citrulline to arginine.

On the basis of such experiments, Beadle proposed a *one gene–one enzyme* hypothesis: that each gene has a single primary function, namely, to control the formation of one, and only one, enzyme. He later modified it to a *one gene–one protein* hypothesis when it became clear that some genes control the formation of proteins that are not enzymes, and it is now usually known as the *one gene–one polypeptide chain* hypothesis since it is clear that many protein molecules are aggregates of more than one kind of polypeptide chain bound by noncovalent bonds.

It is now clear that the hypothesis is, with certain reservations, correct and applies to all living organisms from viruses to man. In the few instances in which two genes appear at first sight to be directly concerned in the formation of one protein an explanation can be found that does not violate the hypothesis. For example, in *E. coli* there is an enzyme, alkaline phosphatase, and mutations in the gene controlling its formation can, as expected, render the bacteria unable to perform the reaction catalyzed by the enzyme. But some of these mutant strains have the ability to catalyze this reaction restored if a second *suppressor* mutation occurs at a quite different gene. It at first appeared that this second gene might also be specifically concerned in the formation of the enzyme, but it is now clear that it is not. The second mutation causes an alteration in the nucleotide sequence of one of the three transfer RNAs that are responsible for incorporating the amino acid tyrosine into polypeptide chains. For reasons that will become evident in the next chapter, this altered transfer RNA enables the suppressed mutant to resume synthesis of alkaline phosphatase.

An important reservation to the one gene–one protein hypothesis has become apparent since it was propounded: certain genes direct the formation of ribosomal and transfer RNA molecules but not the formation of proteins. But this reservation does not mean that genes act in two different ways, for it will be seen that they in fact direct the formation of protein molecules by way of RNA molecules.

2. A Group of Nucleotides to One Amino Acid

Proteins differ from one another in the length and sequence of their amino acid chain or chains. Beadle's hypothesis therefore implied that the DNA of each gene in some way directs the formation of an amino acid chain of a specific length and sequence. If the structure of the DNA is altered by mutation, then either a protein of altered structure or no protein at all will be formed. Experiments will be described in this section that first proved that different mutations in a gene can result in alterations of sequence in different small sections of a protein molecule. This evidence led to precise suggestions as to how the DNA of a gene directs the formation of a protein.

The first evidence for the effect of an alteration in gene structure on the sequence of the corresponding protein came from the work of Sanger and his colleagues, which was discussed in Chapter 2. They found that the sequences of cattle, pig, sheep, horse, and whale insulin were identical except for amino acids 8, 9, and 10 of the chain with N-terminal glycine. Amino acid 8, which is alanine in cattle and sheep insulin, is theonine in that of the pig, horse, and whale. Amino acid 9 is serine in cattle, pig, and whale insulin but glycine in that of the sheep and horse. Amino acid 11 is valine in cattle and sheep insulin but isoleucine in that of the pig, horse, and whale. The most probable explanation for these differences is that, during the evolution of each of these animals from a common ancestor, the gene that determines the amino acid sequence of this chain of insulin has undergone one or more mutations. These mutations have resulted in changes in sequence of a small part of the insulin molecule. The fact that the changes are limited to amino acids 8, 9, and 10 of one chain might suggest that mutations can only alter the sequence in a limited part of a protein. Evidence that this is not so will soon be given. Mutation can certainly change the sequence in other parts of the insulin molecule, but such insulin is defective. Animals that possess it produce fewer offspring than the average, and, so, the mutant gene is eliminated by natural selection.

Comprehensive studies have been made of the effects of mutation on the sequence of the protein hemoglobin of red blood corpuscles. Each molecule of hemoglobin contains four polypeptide chains. Two of these, called the α chains, are identical and each contains 141 amino acids, while the other two, the β chains, are also identical and each contains 146 amino acids. Most people share the sequence of the α chains and of the β chains in common with one another, but a few people have chains of abnormal sequence. The first abnormality was discovered in 1956 by V. M. Ingram at Cambridge University when he investigated the hemoglobin from patients with sickle-cell anemia. In this disease the ability of the hemoglobin to transport oxygen is defective, and it differs slightly in other properties from normal. Ingram showed that this sickle-cell hemoglobin differs from the normal in only one amino acid: the sixth amino acid in the β chains is valine instead of glutamic acid.

That this difference in structure is due to an alteration in the structure of a single gene is confirmed by studying the inheritance of the protein. If a person with sickle-cell anemia marries a normal person, the red blood corpuscles of their children all contain roughly half normal hemoglobin and half sickle-cell hemoglobin. This suggests that they have inherited a normal gene from one parent and a gene of altered structure from the other. This is confirmed by studying the distribution of normal and sickle-cell hemoglobin in the red corpuscles of the grandchildren. Many other kinds of hemoglobin with one abnormal amino acid in the β chain have been isolated from different people, and they are inherited in a similar way. Among the altered amino acids are numbers 6, 7, 26, 63, 63, 67, and 121. Hence, the gene must possess at least seven different abnormalities of structure to cause these alterations in one of the 146 amino acids at different points along the β chain.

Many hemoglobins have also been found that have an altered amino acids in positions along the α chain. Again, these abnormal proteins are inherited in a way that shows that they result from the inheritance of a gene of altered structure. A woman has even been found whose hemoglobin contained both the normal α chain and a roughly equal amount of an abnormal α chain, together with the normal β chain and a roughly equal amount of an abnormal β chain. She married a man with normal hemoglobin, and the distribution of normal and abnormal proteins among their children has shown that different genes on different chromosomes control the sequences of the α and β chains.

H. G. Wittmann in Germany also made a comprehensive study of the effects of mutation on the sequence of the protein formed by tobacco-mosaic virus. It will be remembered that the virus is composed solely of protein and RNA and that the genes of the virus are made of RNA. Wittmann studied the effects of treating the RNA of the virus with nitrous acid before infection. This compound can react with adenine, cytosine, and guanine of the RNA nucleotides, converting their —NH_2 groups to —OH. If the RNA is left only briefly in contact with nitrous acid, so that only one or a few nucleotides per molecule react, it will still infect tobacco leaves. But the lesions that are formed often look different from normal, and from them mutant strains of the virus can be isolated. These strains can be multiplied within the leaves of the plant and their proteins isolated. Wittmann isolated 29 proteins from different viruses in which 1 amino acid somewhere along the chain of 157 was different from normal and 6 proteins in which two amino acids were different.

The most thorough study of the effects of mutation in a gene on the amino acid sequence of the corresponding protein has been made by C. Yanofsky and colleagues of Stanford University. They studied the gene in *E. coli* that controls the formation of one of the two peptide chains (the A chain) of the enzyme tryptophan synthetase. Their work proved a new fact: positions in which a number of different mutations occur along a gene are in the same order as the positions in which the amino acid substitutions they produce occur along the polypeptide chain. Thus they isolated seven *E. coli* mutants that had replacements at amino acids 174, 176, 182, 210, 212, 233, and 234 along the polypeptide chain of 267 amino acids. Mapping experiments showed that the seven changes in structure also occurred in this order along the gene. They also proved that two mutations in different nearby positions in the gene can produce a change in the same amino acid of the protein. Thus two *E. coli* mutants were isolated in which amino acid 47, which is normally glycine, was replaced by arginine in one and by glutamic acid in the other. Crosses between these mutants yielded [a very few] recombinant bacteria with the normal protein, proving that the two mutations were at different [but nearby] positions in the gene.

What may we conclude from all these experiments about the manner in which a gene directs the formation of a protein? Benzer's experiments, described earlier, suggested that a mutation is a change in nucleotide sequence in a limited segment of the DNA of a gene. The facts presented in this section suggest that such changes in nucleotide sequence result in changes in amino acid sequence in a corresponding position along a polypeptide chain. Since mutations at different nearby positions along the gene, as shown by Yanofsky, can affect the same amino acid, it appears that a group of nucleotides within the gene must be responsible for the insertion of one particular amino acid into the polypeptide. This hypothesis will be considered in more detail in the next section.

3. Three Nucleotides to One Amino Acid

Precisely how a group of nucleotides is able to direct a certain amino acid into a protein will be considered in later sections. In this section we shall consider how many nucleotides occur in each group and how the nucleotides of one group are distinguished from those of adjacent groups. A problem that arises is that each gene is composed of the two polynucleotide strands of the double helix. It will be assumed that only one strand directs the formation of a protein. As will be seen later, this assumption has been proved correct.

The simplest version of the problem we are considering would be if each nucleotide in DNA directed one specific amino acid into a protein— a version that, as we have seen, is excluded by Yanofsky's experiments. For example, suppose that the nucleotides containing adenine (A), guanine (G), cytosine (C), and thymine (T) caused the insertion of the amino acids serine, proline, valine, and leucine, respectively. Then, if a DNA strand began with the sequence A–A–G–C–A–T–, it would direct the formation of a protein molecule that began with the sequence ser–ser–pro–val–ser–leu– and contained the same number of amino acids as there were nucleotide pairs in the gene. But it can be seen that, because DNA contains only four kinds of nucleotide, this coding relation could result only in proteins that contained four kinds of amino acid.

It would also be impossible for specific pairs of nucleotides to direct each kind of amino acid because the corresponding protein could contain only 16 amino acids. This is because the nucleotides can be arranged in only 16 combinations: A–A, A–G, A–C, A–T, G–G, G–A, G–C, G–T, C–C, C–A, C–G, C–T, T–T, T–A, T–G, and T–C. However, triplets of nucleotides could direct the 20 amino acids into protein chains because they can be arranged in 64 combinations. Possibly only 20 of these combinations would direct amino acids, but with more than one combination corresponding to each amino acid, possibly more would.

In theory, therefore, the simplest possible coding relation between DNA and protein would be if a succession of nucleotide triplets from one end of the "sense" strand of the DNA of a gene to the other determined the succession of amino acids in a protein. For example, suppose that the gene began at the 5′ end with the sequence A–A–T–G–C–A–C–C–A–. Then the triplet A–A–T– would determine the first amino acid, –G–C–A– the second, and –C–C–A– the third. In this code, the triplets are counted off in threes from the first nucleotide in the DNA; that is, the members of each triplet are determined solely by their position relative to the first nucleotide. This coding relation was thought out by Crick and others. They also devised more ingenious codes in which the members of each triplet are determined in other ways. One suggested that every fourth nucleotide is a "comma." In another, accidentally combined nucleotides of one triplet could not be with those of an adjacent triplet because the sequence of the resulting triplets did not correspond to any amino acid. Overlapping codes were also suggested in which the end of one triplet of nucleotides forms the beginning of the next. For example, in the sequence given above, an overlapping code would be one in which the triplet A–A–T determined the first amino acid, and –A–T–G or –T–G–C– determined the second. However, experiments (which will be described in this section) of F. H. C. Crick, S. Brenner, and their colleagues around 1960 proved that the code that has arisen in nature is the simplest one.

Evidence was first needed that the code is not overlapping, and this came from the experiments of Wittmann on tobacco-mosaic virus, which were described in the last section. By chemical treatment he altered isolated nucleotides in the RNA of the virus and found that usually only one amino acid in the protein became changed. If two amino acids were changed, they were never adjacent. This would not be found in an overlapping code in which a nucleotide in one group can also form part of an adjacent group and so be concerned in directing more than one amino acid. It could also be concluded that the code in human beings is not overlapping because all abnormal hemoglobins have changed in only one amino acid.

Crick and Brenner's experiments suggested that groups of three nucleotides direct each amino acid and that the members of each group are determined by their position relative to the first nucleotide of the gene. They performed their experiments on the T_4 bacterial virus and its rII mutants using the techniques developed by Benzer, which have already been described. It will be remembered that rII mutants of the virus will not grow on $E.$ $coli$ of strain K and that they produce colonies of unusual appearance when grown, in agar jelly, on $E.$ $coli$ B. It will also be remembered that although these rII mutants look alike, they had been proved by Benzer to be of many kinds, each with a structural change in a different limited segment of the gene. Crick and Brenner's experiments were founded on the discovery by Benzer that, by double infection of $E.$ $coli$, the structural defects in the genes of two viruses can be recombined to give progeny that contain both these defects in one gene.

The interpretation of these experiments involved a number of assumptions that were justified by the way in which they produced clarity out of what would otherwise have been a chaos of observations. Crick and colleagues selected about 80 distinct rII mutants of the T_4 virus that, it was concluded, had one nucleotide missing from their DNA or one extra nucleotide inserted into it at some point. The gene of each virus was, therefore, one nucleotide longer or shorter than normal. Some of the viruses had been produced by the action of acridines, which were believed to cause mutations in this way, but the conclusion that a nucleotide had been inserted or removed was largely founded on the following property of the viruses. It was found that they could be divided into two groups. If the structural defects in any two viruses of one group were combined into one gene, the progeny were, like their parents, all *r*II mutants that would not grow on *E. coli* K. But if the defects in viruses of the different groups were combined, progeny were often produced that showed a surprising property. They would grow on *E. coli* K and formed colonies on *E. coli* B that *more or less* resembled those of the normal virus. In *r*II mutants, it is assumed that an unidentified protein is usually lacking, owing to damage of the gene. Crick and colleagues concluded that the viruses that would now grow on *E. coli* K contained a protein that *more or less* resembled the normal one. This protein was formed because the gene had been brought back to its normal length by combining the region lacking a nucleotide in a virus of one group with the region containing an extra nucleotide in a virus of the other group. The reason this could result in the formation of a protein of more or less normal structure is as follows.

It was assumed that the nucleotides that fall into a single group or *codon* are determined solely by their position relative to the first nucleotide of the gene. It was on this assumption that the argument was founded. Suppose, for simplicity, that the normal protein, which is lacking in the *r*II mutants, contained only one kind of amino acid that is directed into the protein by the group of nucleotides –T–A–G– (the size of the group is immaterial to the argument but has been assumed to be three). Then the sense strand of the gene will consist merely of this sequence repeated throughout its length, with three times as many nucleotides as there are amino acids in the protein:

T– A– G– T– A– G– T– A– G– T– A– G– T– A– G– and so on.
1   2   3   4   5   6   7   8   9   10 11 12 13 14 15

Suppose also that one *r*II virus lacked the fourth nucleotide in this gene and that another had an extra nucleotide (A) inserted between the seventh and eighth. Then the sequence of these genes would be

T– A– G– A– G– T– A– G– T– A– G– T– A– G– T– and so on,
1   2   3   5   6   7   8   9   10 11 12 13 14 15 16

T– A– G– T– A– G– T– A– A– G– T– A– G– T– A– and so on.
1   2   3   4   5   6   7   8   9   10 11 12 13 14

In each of these abnormal genes the nucleotides have been numbered according to the position they occupied in the normal gene. In each gene the first member of each group of three nucelotides has been shown in boldface type. It will be seen that in the abnormal genes, after the point of deletion or insertion of a nucleotide, the triplets have become A–G–T and G–T–A, respectively. Hence incorrect amino acids, or no amino acids at all, will be inserted by these triplets into the protein that, if formed, will have a structure very different from normal. Therefore, viruses with either of these genes would be expected to be rII mutants.

However, if by double infection nucleotides 1 through 5 of the gene of the first virus were combined with nucleotides 6 onwards of the second, a virus would be formed whose DNA had the original length and the sequence

T– A– G– A– G– T– A– A– G– T– A– G– T– A– G– and so on.
1 2 3 5 6 7   8 9 10 11 12 13 14 15

It can be seen that, after the second alteration, the correct triplet T–A–G is restored. Provided that the incorrect triplets do correspond to some amino acid, a protein will be formed that has only incorrect amino acids along a limited segment of the molecule. If the deletion and insertion of nucleotides are not too far apart, it is reasonable to assume that the sequence of the protein that is formed will be normal enough to allow the virus to grow on E. coli K. In support of this, it was found that the viruses would not infect E. coli K if the deletion and the insertion were more than a certain distance apart on the gene (their positions having been found by the mapping techniques of Benzer).

The experimental observations of Crick and his colleagues could, therefore, be explained if it was assumed that the nucleotides that fall together into a single functional group are determined solely by their position relative to the first nucleotide of the gene. Hence the experiments provided some support for these assumptions. But the argument was tenuous and may appear unconvincing. However, it was supported by a further discovery that made the experiments difficult to interpret in any other way and that strongly suggested that there are three nucleotides to each codon. It was mentioned that if viruses that had two nucleotides more or less than normal in the gene were produced by recombination, they were always rII mutants. The brilliant discovery was made that if viruses that had three nucleotides more or less than normal were formed by further recombination, they would often grow on E. coli K and form more or less normal colonies on E. coli B. This can be explained by combining the previous assumptions with the assumption that there are three nucleotides to each functional group. Suppose that three rII viruses lack the fourth, eighth, and tenth nucleotide, respectively. The sequence of their genes will be

T– A– G– A– G– T– A– G– T– A– G– T– A– G– T– and so on,
1 2 3 5 6 7 8 9 10 11 12 13 14 15 16

T– A– G– T– A– G– T– G– A– T– G– A– T– G– A– and so on,
1 2 3 4 5 6 7 9 10 11 12 13 14 15 16

T– A– G– T– A– G– T– A– G– A– G– T– A– G– T– and so on.
1 2 3 4 5 6 7 8 9 11 12 13 14 15 16

After each deletion, incorrect triplets are repeated over the whole length of
the genes. Suppose, now, that the three deletions are combined into one gene.
Its sequence would be

T– A– G– A– G– T– G– A– G– T– A– G– T– A– G– and so on.
1 2 3 5 6 7 9 11 12 13 14 15 16 17 18

It can be seen that after the third deletion the correct triplet is restored for
the remainder of the gene. Hence the gene might be expected to form a
protein normal enough in sequence to allow growth on *E. coli* K. The same
conclusion is reached if three insertions are combined or if the sequence of
the DNA and the protein is more complex than in this simple example.

4. Messenger RNA

We have concluded that every three nucleotides along one of the paired DNA
strands of a gene directs the insertion of one amino acid into the molecule
of the corresponding protein. It is now logical to ask, Which triplet of
nucleotides corresponds to each of the 20 amino acids that can appear in a
protein? But it is not possible to answer this question until a new fact has
been introduced: the nucleotide sequence of the sense strand of a gene must
first be translated into that of a molecule of *messenger RNA* before synthesis
of a protein. Hence it is triplets of nucleotides along the messenger RNA
that in fact direct the insertion of each amino acid.

That most proteins are not formed in contact with genes in the nucleus
of an animal cell but rather in the cytoplasm in contact with ribosomes was
first proved in the 1950s by H. Borsook and colleagues at the California
Institute of Technology and P. C. Zamecnik and colleagues at Harvard
University. Borsook realized that the way to find at which point amino acids
are incorporated into proteins, in the cells of higher animals, was to inject a
radioactive amino acid into an animal and kill it a few minutes after the
injection. After this short time, the only proteins to contain large amounts
of the radioactive amino acid would probably be those at the site of protein
formation. He therefore injected a radioactive amino acid into a guinea pig.
Thirty minutes later he killed it and disintegrated its liver in a sucrose
solution. He centrifuged the resulting suspension at three different speeds and

obtained the liver cell nuclei, the mitochondria, the microsomes, and the supernatant liquid. He discovered that the protein of the microsomes had over twice as much radioactivity per gram as the protein of any of the other fractions. The microsomes therefore seemed to be the point in the cell at which amino acids are converted into proteins, and other experiments confirmed this.

In 1950, when Borsook did this experiment, it was not known that microsomes are not true cell structures but rather fragments of the endoplasmic reticulum to which ribosomes are attached. When this fact was discovered a few years later, P. C. Zamecnik and his colleagues attempted to find whether proteins are built up on the ribosomes or on the remainder of the endoplasmic reticulum. In each of a series of experiments, they anesthetized a rat, opened its abdomen, and exposed its liver. They then injected a solution of a radioactive amino acid into its tail vein. Then, between 2 and 20 minutes later, they rapidly removed part of the liver and plunged it into ice. They then separated the microsomes and washed them with a solution of a detergent that frees the ribosomes, which they then separated by centrifuging. They found that, 2 minutes after injecting the radioactive amino acid, the ribosomes contained up to seven times more radioactivity in every milligram of protein than the remainder of the microsomes. This difference in radioactivity persisted for several minutes after the injection. It seemed clear that proteins were formed on the ribosomes. Later experiments have fully supported this conclusion and have shown that proteins are also formed on ribosomes in other organisms including bacteria.

Because proteins are formed on ribosomes rather than directly on the genes, it follows that each ribosome must carry some kind of precise copy of a gene, in contact with which amino acids are assembled into proteins of the correct sequence. Since over half the weight of a ribosome is RNA, it at first seemed probable that this RNA would carry an imprint of the gene by its sequence being directly related to that of the DNA. It will be remembered that when DNA forms self-copies, the paired strands separate and new strands are formed beside them according to the following rule: *Where nucleotides with the bases A, G, C, or T occur in one strand, then those with T, C, G, and A, respectively, will occur in the new strand.* This process results from the hydrogen bonding between the bases A and T and the bases G and C. It will also be remembered that the common nucleotides of RNA are very similar to those of DNA, except that one of the four kinds carries the base uracil (U) instead of thymine. It can be proved, by building molecular models, that a nucleotide with the base A will as readily form hydrogen bonds with a nucleotide with the base U as with one with the base T. Therefore it seemed reasonable that the RNA of each ribosome would consist of single or paired molecules, built up in contact with one or both strands of the DNA of a gene according to the following rule: *Where nucleotides with the bases A, G, C, or T occur in a DNA strand, those with U, C, G, and A, respectively, will occur in the RNA strand.* For example, if the sequence of part of the DNA strand happened to be –A–A–G–C–T–, that of the corresponding part of the RNA molecule would be –Ո–Ո–Ɔ–Ɔ–∀–.

The theory that the RNA of the ribosomes is related to the DNA of the genes in this way can easily be tested in bacteria. In bacterial cells, unlike the cells of animal and plant tissues in which many genes must be inactive, almost every gene of every cell must have, on a ribosome, a replica of itself that is directing the formation of proteins. If the theory is correct, the ratio $(G+C)/(A+T)$ in the DNA of each species of bacteria should, at least roughly, equal the ratio $(G+C)/(A+U)$ in the ribosomal RNA. In fact, it does not, as was first discovered by A. N. Belozersky of Moscow State University. He and his colleagues determined these ratios in the total DNA and RNA of 21 species of bacteria. The ratio of $(G+C)/(A+T)$ in the DNA ranged from 0.45 to 2.73. That of $(G+C)/(A+U)$ in the RNA ranged only from 1.03 to 1.45. The bulk of the RNA in these bacteria is ribosomal RNA. If the RNA of the ribosomes themselves is analyzed, the correspondence is no closer.

The theory can be saved only if it is assumed that only a small part of the ribosomal RNA is a replica of the genetic DNA, and this, in fact, is the situation. This RNA has been named messenger RNA because it transports a replica of the nucleotide sequence of the genes into the cytoplasm. It was first detected in 1957 by E. Volkin and L. Astrachan of Oak Ridge National Laboratory in *E. coli* bacteria that had been infected with T_2 or T_7 viruses, in the following experiment.

The bacteria were grown in an aerated culture solution and, at the height of growth, a suspension of one of the viruses was added, followed immediately by a solution of sodium orthophosphate containing radioactive ^{32}P. Then, at short intervals, samples of the infected bacteria were taken, and the RNA was isolated. The quantity of RNA did not increase perceptibly after infection, but all four of the component nucleotides acquired radioactive phosphorus, showing that a small quantity of RNA had, in fact, been formed. After infection with T_2 virus, the quantity of radioactive phosphorus acquired by the nucleotides containing A and U averaged about 1.7 times that acquired by those containing $G+C$. It may be concluded that the ratio $(A+U)/(G+C)$ in the newly formed RNA was roughly 1.7. The ratio in the RNA formed after infection with the T_7 virus was much lower, namely, 1.2. The ratio $(A+T)/(G+C)$ in the DNA of the T_2 and T_7 viruses was found by chemical analysis to be 1.87 and 1.11, respectively. There was, therefore, good evidence that after infection RNA is formed within the bacteria whose sequence replicates that of one or both of the paired DNA molecules of the infecting virus.

A few years later messenger RNA of T_4 bacteriophage was isolated from infected *E. coli* by hybridization with T_4 DNA (see p. 76). Its ratio $(A+U)/$ $(G+C)$ was 1.8 and equal to the ratio $(A+T)/(G+C)$ of the phage DNA. However, the phage RNA contained 21% G and 15% C, unlike the DNA in which the percentages of G and C were equal. It was concluded that messenger RNA is formed on only one strand of the DNA double helix, and this conclusion has been confirmed by other experiments. The messenger RNA formed in *E. coli* after virus infection was proved to bind to bacterial ribosomes, whereupon proteins peculiar to the virus were synthesized. It was also shown that when *E. coli* grow normally their own DNA continually forms messenger RNA, which becomes attached to the ribosomes, the quantity of messenger being small compared with the remainder of the ribosomal RNA and being continually degraded and replaced by fresh messenger. The manner in which ribosomes become attached to the messenger RNA in both microbial and mammalian cells has since been made clear by electron microscopy: a small number of ribosomes are strung along the molecule of messenger RNA to form a *polysome* (Figure 29).

Figure 29. Polyribosomes. The black blobs are *E. coli* ribosomes joined by threads of messenger RNA which run north-south in the picture. The two threads running east-west are parts of the *E. coli* chromosome. The messenger RNA threads are still attached to the chromosome, presumably by RNA polymerase molecules, and the RNA threads increase in length from left to right because the RNA polymerase moves in this direction along the DNA. The picture confirms other evidence that in bacteria ribosomes bind to messenger RNA and start protein synthesis while the RNA is still being synthesized. The growing peptide chain presumably attached to each ribosome is not visible. (Electron micrograph courtesy of Dr. O. L. Miller, Jr., and Dr. Barbara A. Hamkalo, Biology Division, Oak Ridge National Laboratory, U.S.A.)

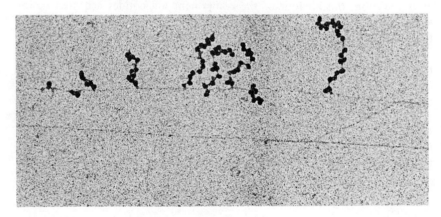

The enzyme that catalyzes the formation of messenger and other RNA has been isolated from several organisms and named RNA polymerase. It resembles Kornberg's DNA polymerase in that it requires DNA as a template. When the four nucleoside triphosphates of RNA are incubated with the enzyme and template DNA, RNA is formed whose sequence is complementary to one or both strands of the template (Figure 30). The RNA chain is built up, starting at its 5' end, by reactions similar to those shown in Figure 26 (p. 79).

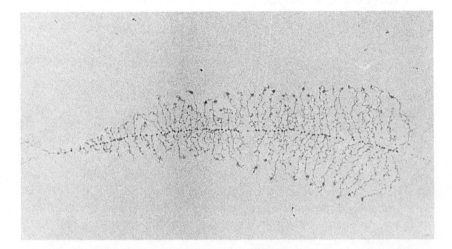

Figure 30. RNA polymerase in action. The strands that cross the picture are the part of the chromatin of a newt (*Triturus viridescens*) that contains the genes for ribosomal RNA. These genes are repeated many times along each strand. The fuzzy threads attached to the strands are ribosomal RNA molecules of gradually increasing size, each attached to an RNA polymerase molecule on the chromatin. (Electron micrograph courtesy of Dr. O. L. Miller, Jr., and Dr. Barbara Beatty, Biology Division, Oak Ridge National Laboratory, U.S.A.)

5. Breaking the Code

It is therefore clear that messenger RNA on polysomes guides amino acids into proteins of the correct sequence and that each molecule of messenger RNA has the same number of nucleotides as the DNA on which it was formed, and a complementary nucleotide sequence to one of the strands. It may be concluded from this, and from experiments discussed earlier in this chapter, that each group of three nucleotides along a molecule of messenger RNA directs one amino acid into a corresponding position in a protein molecule. In this section, experiments will be discussed that show which particular triplet, or triplets, of nucleotides directs each kind of amino acid into a protein. Is phenylalanine, for example, directed by –U–U–U– or by one or more of the other 64 combinations? This breaking of the code resulted from a discovery of M. W. Nirenberg and J. H. Matthaei of the National Institutes of Health. They found that, in extracts of E. coli, the natural messenger RNA could be replaced by synthetic RNA of known sequence. When this was done, abnormal proteins were formed, the sequence of which revealed the code relation of messenger RNA to protein.

Nirenberg and Matthaei made their discovery while studying the formation of proteins by extracts of E. coli. These extracts were prepared as follows: a suspension of the bacteria, which were rapidly growing, was cooled and centrifuged. The bacteria in the sediment were then broken by grinding with aluminium oxide powder. A buffer solution was added, and the mixture was centrifuged at 20,000 times the force of gravity to remove the aluminium oxide and unbroken bacteria. To this bacterial extract was added a small quantity of an enzyme DNAase, which breaks DNA into small fragments. The extract was further centrifuged to remove solid matter, and the supernatant, which still contained the ribosomes, was dialyzed—a process that removes molecules smaller than those of proteins and nucleic acids. It was found, as expected from previous work, that when certain compounds were added to this extract, it would incorporate amino acids into proteins. The formation of proteins was detected by adding radioactive amino acids and, after incubation, precipitating the proteins and measuring their radioactivity. However, after about 20 minutes of incubation, the extracts lost the ability to form proteins.

Nirenberg and Matthaei argued that the reason protein synthesis ceased was because messenger RNA was destroyed after 20 minutes and no more could be formed. They then made a beautifully simple and classic experiment of molecular biology. Could not the natural messenger RNA, which was being destroyed, be replaced by synthetic RNA of known sequence? To an extract that had lost the ability to form proteins after 20 minutes of incubation, they added 10μg of polyuridylic acid (poly(U))—a synthetic RNA containing only nucleotides with the base uracil. When this was incubated with any radioactive amino acid except phenylalanine, little or no radioactive protein was formed. But when incubated with radioactive phenylalanine, the protein precipitate was highly radioactive, and it was proved to contain only one radioactive "protein," namely, polyphenylalanine. Since three nucleotides of messenger RNA appear to direct the insertion of one amino acid into a protein, they concluded that the insertion of phenylalanine is determined by three nucleotides each containing uracil. It was therefore possible to deduce the course of events when a gene directs the formation of a protein containing phenylalanine. The paired DNA strands of the gene each contain three times as many nucleotides as the protein contains amino acids. One of these strands has the sequence –A–A–A– at positions along the nucleotide chain that correspond to the positions of phenylalanine along the protein chain. Under the influence of the enzyme RNA polymerase, messenger RNA is formed in contact with this DNA and contains the triplet –U–U–U– wherever the DNA contained –A–A–A–. The triplet –U–U–U– somehow directs phenylalanine into the protein. Hence the code for phenylalanine is –A–A–A– on the sense strand of the gene and –U–U–U– on the messenger RNA.

Since 1961 when Nirenberg and Matthaei announced their discovery, the genetic code has been completely broken by experiments of two kinds. The first has been an extension of their original experiment but with other synthetic messengers, the most important being RNA molecules of known sequence synthesized by H. G. Khorana and his colleagues at the University of Wisconsin. One of these was poly(A–A–G), that is, RNA with the sequence A–A–G–A–A–G–A–A–G–A–A–G–, etc. When this was added to the *E. coli* extracts, three amino acids were polymerized: lysine to polylysine, arginine to polyarginine, and glutamic acid to polyglutamic acid. Hence each of the three triplets along this RNA (i.e., A–A–G, A–G–A, and G–A–A) must be the codon for one of these amino acids.

The second type of experiment by which codons were assigned merely required the use of trinucleotides such as –U–U–U. A solution of a trinucleotide was added to a suspension of ribosomes. A radioactive amino acid linked to its transfer RNA (see the next section) was then added. If the trinucleotide was the codon for this amino acid, then the ribosomes would bind both the trinucleotide and its aminoacyl transfer RNA in an abortive attempt at protein synthesis. The binding could be detected by filtering off the ribosomes and measuring their radioactivity. This method revealed the order of different nucleotides in a triplet, and which lies at the 5′ and which at the 3′ end. The codons that have been assigned by these methods are shown in Figure 31.

That this code is in fact the one used by the living cell is verified by changes in amino acid sequence in mutated proteins. The most probable origin of replacements of one amino acid by another is errors in DNA replication that replace one base pair by another—the chance of two adjacent replacements occurring is very small. The many amino acid replacements that have been found in human hemoglobins can all, in fact, be attributed to changes in one base in the codons of Figure 31. Thus in sickle-cell hemoglobin a glutamic acid residue (codons G–A–A and G–A–G) is replaced by valine (codons G–U–A and G–U–G). This also suggests that the code is the same in man as in *E. coli*.

A coding problem that remains is the following. We have seen that U–U–U and A–A–A are codons for phenylalanine and lysine respectively. If the sequence Phe-Lys (i.e. Phe is N-terminal) is inserted in a protein by these codons, are they arranged on the messenger RNA in the order

U–U–U	Phe	U–C–U	Ser	U–A–U	Tyr	U–G–U	Cys
U–U–C	Phe	U–C–C	Ser	U–A–C	Tyr	U–G–C	Cys
U–U–A	Leu	U–C–A	Ser	U–A–A	Chain termination	U–G–A	Chain termination
U–U–G	Leu	U–C–G	Ser	U–A–G	Chain termination	U–G–G	Try
C–U–U	Leu	C–C–U	Pro	C–A–U	His	C–G–U	Arg
C–U–C	Leu	C–C–C	Pro	C–A–C	His	C–C–C	Arg
C–U–A	Leu	C–C–A	Pro	C–A–A	Gln	C–G–A	Arg
C–U–G	Leu	C–C–G	Pro	C–A–G	Gln	C–G–G	Arg
A–U–U	Ile	A–C–U	Thr	A–A–U	Asn	A–G–U	Ser
A–U–C	Ile	A–C–C	Thr	A–A–C	Asn	A–G–C	Ser
A–U–A	Ile	A–C–A	Thr	A–A–A	Lys	A–G–A	Arg
A–U–G	Met	A–C–G	Thr	A–A–G	Lys	A–G–G	Arg
G–U–U	Val	G–C–U	Ala	G–A–U	Asp	G–G–U	Gly
G–U–C	Val	G–C–C	Ala	G–A–C	Asp	G–G–C	Gly
G–U–A	Val	G–C–A	Ala	G–A–A	Glu	G–G–A	Gly
G–U–G	Val	G–C–G	Ala	G–A–G	Glu	G–G–G	Gly

Figure 31. Genetic code: triplets of nucleotides in messenger RNA that bind transfer RNAs for the amino acids shown.

–U–U–U–A–A–A– or –A–A–A–U–U–U–, where the hyphen on the left of each sequence represents its 5′ end and that on the right its 3′ end? In other words, does the N-terminus of an amino acid sequence correspond to the 5′ or the 3′ end of messenger RNA? Evidence that the 5′ end corresponds to the N-terminus first came from the demonstration that poly(A) with a single C at its 3′ end directs the formation of polylysine with a single asparagine at its C-terminus. The same conclusion was reached from study of a lysozyme produced by a mutant of the bacterial virus T_4. The protein formed by normal T_4 has the sequence

–Lys–Ser–Pro–Ser–Leu–Asn–Ala–.

In the mutant this is changed to

–Lys–Val–His–His–Leu–Met–Ala–.

This change can be attributed to the deletion (removal) of one base pair in the DNA and insertion of another at a different point, so changing the corresponding sequence in messenger RNA from

–A–A–A–A–G–U–C–C–A–U–C–A–C–U–U–A–A–U–G–C–U–

to

–A–A–A–G–U–C–C–A–U–C–A–C–U–U–A–A–U–G–G–C–U–.

This also confirms that bacteriophage T_4 uses the same codons as *E. coli* and man.

In 1972 the complete nucleotide sequence of a gene, and the complete amino acid sequence of the protein that it directs, were reported. The gene directs the formation of the coat protein of the bacterial virus MS2, whose chromosome is a single strand of RNA which also acts as messenger for synthesis of the virus proteins. The coat protein has 129 amino acids and hence 387 nucleotides in the RNA direct their insertion. Once again, all amino acids are inserted by codons in Figure 31, and 48 of them are used, there being interesting omissions. For example, while all four codons are used to insert the 14 residues of valine, only one of the two codons is used for the 4 residues of tyrosine.

6. Transfer RNA

In this section we shall consider precisely how the nucleotides of messenger RNA guide amino acids into proteins of the correct sequence. How, for example, do three nucleotides with the bases U–U–U cause the insertion of phenylalanine into a protein? We know that the messenger RNA becomes attached to the ribosomes. Do the amino acids assemble in a row on the surface of the messenger RNA and link together to form a protein under the influence of an enzyme? Does the triplet U–U–U have some particular shape against which phenylalanine molecules fit neatly, while those of the other amino acids do not, and do the other triplets each have a shape against which only one of the other 19 amino acids will fit neatly? Such a mechanism seems improbable. The actual mechanism is far more elegant—and was predicted in outline by F. H. C. Crick.

What molecule, in particular, will a triplet of nucleotides with, say, the bases U–U–U attract? The obvious answer is a nucleic acid molecule containing a triplet of nucleotides with the bases A–A–A in an exposed position, which would be held by hydrogen bonding. If phenylalanine was attached to such a molecule, it would therefore become held on the correct region of the messenger RNA. Suppose that the adjacent triplet of nucleotides on the messenger RNA carried the bases C–C–C which is a codon for proline. This would attract a nucleic acid molecule with the bases G–G–G in an exposed position. If proline was attached to such a molecule, it would be held on the messenger RNA in the correct position beside phenylalanine. Evidence will be given in this section that RNA molecules of this kind, which are specific for different amino acids, do exist in free solution in living cells and are responsible for the transport of amino acids to ribosomes. These molecules are known as transfer RNA. In effect each acts as an adaptor that gives an amino acid the right shape for fitting in the correct position on the messenger RNA of the ribosomes. The amino acids, which become arranged in this way in the correct order, are then linked together into a protein of the correct sequence under the influence of a number of enzymes.

Transfer RNA, and also the reactions by which amino acids become attached to it, were largely discovered by M. B. Hoagland at Harvard University. Each amino acid first reacts with adenosine triphosphate (ATP), catalyzed by an enzyme—the "activating enzyme"—that is specific for that amino acid. The amino acid becomes attached by its carboxyl group to the innermost of the three phosphoric acid groups of ATP, the other two being eliminated as pyrophosphate. The compound that is formed has the structure shown in Figure 32. This compound never becomes separated from the enzyme but immediately reacts with a molecule of the transfer RNA that is specific for the particular amino acid, the reaction being catalyzed by the same enzyme as before. In this reaction the amino acid is removed from the adenosine monophosphate and becomes bonded to the transfer RNA, and the enzyme is released. All kinds of transfer RNA have a nucleotide containing adenine at the end of the molecule with a free 3'-hydroxyl (Figure 18, p. 45). The amino acid becomes linked to this nucleotide by a bond between its carboxyl group and the 3'-hydroxyl group of the ribose.

An experiment will now be described that in 1962 first proved that

Figure 32. Activation of an amino acid by reaction with ATP.

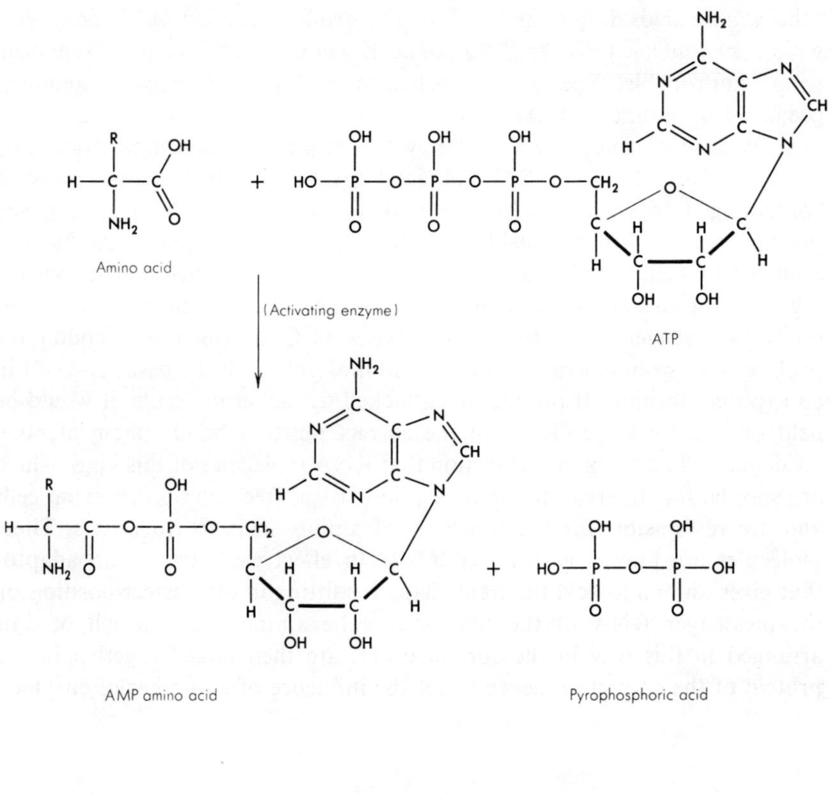

it is the transfer RNA that seeks out the correct triplet on messenger RNA. It was done jointly by workers at Rockefeller University, Johns Hopkins University, and Purdue University and shows that the transfer RNA is solely responsible for finding the correct triplet and that the amino acid is merely carried by it as a passenger.

In this experiment the amino acid, cysteine, was linked to its specific transfer RNA and was then converted to the amino acid alanine, without altering the transfer RNA. It was found that the transfer RNA still behaved in protein synthesis as if it were attached to cysteine. The experiment was performed as follows.

A mixture of the various kinds of transfer RNA was isolated from *E. coli* bacteria. *E. coli* cells were also disintegrated in buffer and centrifuged at 100,000 times gravity to give an extract that contained the activating enzymes that catalyze the attachment of amino acids to transfer RNA. This extract was incubated with the transfer RNA mixture and some of the amino acid cysteine, which contained radioactive carbon, together with ATP and certain other compounds. The cysteine became attached to its specific transfer RNA, and the resulting compound was isolated. It was then shaken in a buffered solution for 30 minutes with Raney nickel—a suspension of nickel that catalyzes the reduction of compounds. The radioactive cysteine became reduced to radioactive alanine, but this remained attached to the RNA responsible for the transfer of cysteine.

Now, Nirenberg and his colleagues had shown that synthetic messenger RNA composed of nucleotides with the bases U and G directs cysteine into proteins. Alanine is not directed by this particular messenger RNA. The compound between radioactive alanine and the transfer RNA specific for cysteine was therefore incubated with poly(U–G) and the extracts of *E. coli* used by Nirenberg. The protein was then precipitated with trichloracetic acid. It was found to contain large amounts of radioactivity, showing that the alanine had been incorporated into protein by the specific messenger RNA for cysteine. It is clear from this experiment that transfer RNA contains all the chemical groups necessary for attachment to the correct triplet in messenger RNA and that the amino acid is not involved.

Since that time a number of kinds of transfer RNA have been isolated and their complete nucleotide sequences determined. That of alanine transfer RNA was shown in Figure 18, p. 45. The looping of the nucleotide chain in this figure into a "cloverleaf" structure by cross-linking of hydrogen bonds is hypothetical. Nevertheless, it is almost certainly correct since all transfer RNAs whose sequence is known can be arranged into a structure of this type. Moreover, when so arranged, the *anticodon* or nucleotide triplet that will hydrogen-bond with the codon for the particular amino acid on messenger RNA is always found in the central loop. Also, certain sequences in the other two loops are common to all transfer RNAs, suggesting that they have a common function in binding the transfer RNA to the ribosomes.

The triplet on the central loop of alanine transfer RNA, which is the alanine anticodon, is I–G–C. I represents the base inosine, which is formed by replacement of the amino group on position 6 of adenine by hydroxyl after synthesis of the transfer RNA. Like adenine, inosine will hydrogen-bond with uracil. Alanine has four codons (Figure 31): G–C–U, G–C–C, G–C–A, and G–C–G. Hence the anticodon I–G–C is complementary to the codon G–C–U—remembering that the pairing of complementary nucleotide chains is antiparallel (p. 69):

$$-\text{C}\text{-}\text{C}\text{-}\text{I}-$$
$$-\text{G}\text{-}\text{C}\text{-}\text{U}-$$

Do each of the three remaining codons for alanine pair with a different transfer RNA? The answer to this question illustrates some important points about codon: anticodon pairing. First, it has become clear that codons in messenger RNA that differ only in their third nucleotide do often pair with the same anticodon. In the present example G–C–C and G–C–A as well as G–C–U pair with the anticodon I–G–C of the alanine transfer RNA of Figure 18, p. 45. This appears to be because transfer RNA molecules are so folded that the first nucleotide of the anticodon (i.e., the one at the 5′ end) has more freedom of movement ("wobble") than the other two, and hence its base is able to complement a wider range of bases on the messenger RNA.

This so-called wobble theory predicted correctly that whenever the first base in the anticodon of transfer RNA is G it will pair with both U and C, whenever it is U it will pair with both A and G, and whenever it is I it will pair with A, U, and C. These facts show that many of the different codons for a single amino acid (Figure 31) bind to the same transfer RNA. They also show that some amino acids must have more than one transfer RNA. Thus there must be another alanine transfer RNA to pair with the codon G–C–G. When an amino acid has more than one transfer RNA, the same activating enzyme has been found to catalyze the attachment of the amino acid to each.

There is one further complicating feature about transfer RNAs: some amino acids, of which one is tyrosine, have two transfer RNAs whose nucleotide sequence differs but not at the anticodon. Hence, both pair with the same codon.

7. Protein Synthesis

We can now outline our knowledge of how amino acids are polymerized into polypeptide chains of the correct sequence and length. Each of these polypeptides has an amino acid with its amino group free at one end of the chain (the N-terminus) and an amino acid with its carboxyl group free at the other end (the C-terminus). We must first know in what order in time the amino acids are incorporated into the polypeptide during its synthesis. Is the amino acid that will form the N-terminus first linked by its carboxyl to the amino group of its neighbor and the chain then extended one at a time with the C-terminal amino acid added last? Or is it built up in the opposite direction with the C-terminal amino acid being first linked to its neighbor? Or are different regions of the chain built up separately and later joined?

The answer to this question has come from isotopic tracer experiments of which we shall describe the one published by D. N. Luck and J. M. Barry— although the race for the correct answer was in fact won by R. Schweet and colleagues with H. M. Dintzis second. In this experiment we collected pancreases from cattle at the Oxford slaughter yard, brought them back to the laboratory on ice, and with a razor blade cut slices that were slightly thicker than the average pancreatic cell. These slices we incubated at 37°C in a phosphate buffer with added glucose and the amino acid lysine labeled with ^{14}C. The pancreas synthesizes the digestive enzyme ribonuclease, and from our slices we isolated ribonuclease that had incorporated the labeled lysine.

We then hydrolyzed the ribonuclease with the enzyme trypsin. Trypsin severs protein chains wherever the carboxyl group of lysine or arginine is linked to an adjacent amino acid. Hence, on hydrolysis of ribonuclease with trypsin the radioactive lysine was distributed among the different peptide fragments. We isolated each of these peptides by chromatography on an ion-exchange resin. We then hydrolyzed each with hydrochloric acid, isolated its lysine by chromatography, and measured its specific radioactivity. The complete amino acid sequence of ribonuclease had been determined a few years earlier by S. Moore and W. H. Stein at Rockefeller University (Figure 13, p. 39). By reading their experimental details, we were able to deduce from which position in the protein chain each of our eight samples of lysine was derived.

When the pancreas slices had been incubated for 2 hours with the labeled lysine, each lysine residue of ribonuclease had the same high specific radioactivity. But after only 3 minutes of incubation the specific radio-activities were low and, moreover, differed from one another in a beautifully ordered way. The residue at the N-terminus was the least radioactive and that nearest the C-terminus was the most active; the remaining six residues had intermediate radioactivities, and these increased with increasing distance from the N-terminus.

These results show that the ribonuclease molecule is built up by a series of reactions in which the N-terminal amino acid is first linked to its neighbor, which is in turn linked to its neighbor, and so on, until the molecule is completed by the addition of the amino acid at the C-terminus. Only this mechanism is consistent with the continuous increase in the specific radioactivity of the lysine residues from the N-terminus, which can be explained as follows. The tissue slices at the moment of addition to the medium with radioactive lysine contained many incomplete ribonuclease molecules of varying length. All contained the lysine at the N-terminus, but progressively fewer contained the lysine residues that lie at increasing distances toward the C-terminus. Hence, when the ribonuclease molecules were completed, progressively more molecules of radioactive lysine were incorporated into the positions lying at increasing distances from the N-terminus. Ribonuclease molecules formed subsequently would have radioactive lysine at all positions along the chain, and therefore after 2 hours of incubation these evenly labeled molecules would mask the unevenly labeled molecules formed in the first few minutes. Similar experiments have also shown that molecules of hemoglobin and egg white lysozyme are built up from the N-terminus to the C-terminus. We have seen that messenger RNA is synthesized from its 5′ end (p. 103) and that this end corresponds to the N-terminus of the protein (p. 106). Hence messenger RNA might initiate protein synthesis at its 5′ end before its own synthesis was complete, and in bacteria it does so. (See Figure 29, p. 102.)

Protein synthesis involves messenger RNA, ribosomes, and up to 20 kinds of amino acid, each linked to its own transfer RNA. The addition of successive amino acids to a protein chain occurs at a ribosome that moves along the messenger RNA from the codon for the N-terminal to that for the C-terminal amino acid. Electron micrographs usually show more than one ribosome attached to a molecule of messenger RNA to form a *polysome*, because more than one ribosome at a time is able to move down the messenger RNA with a growing polypeptide chain attached.

A ribosome is composed of a large subunit and a smaller one of about half its size, and when free in the cytoplasm of the cell, ribosomes dissociate into these subunits. The formation of a protein chain begins by a small ribosomal subunit binding to a molecule of messenger RNA near the codon for the N-terminal amino acid of a protein. The correct transfer RNA with the N-terminal amino acid attached also binds to the codon for the N-terminal amino acid at the same time. In *E. coli* the N-terminal amino acid is always formylmethionine, that is, the normal amino acid methionine with its amino group formylated. After synthesis of the protein, the formyl group, and sometimes the N-terminal methionine also, is removed. Formyl-methionine is incorporated only at the beginning of a protein chain, and the formyl group would in fact prevent its being incorporated at any other position. Formylmethionine is formed by the formylation of methionine after it has become linked to one of two kinds of methionine transfer RNA. Both these transfer RNAs bind to the same codon on messenger RNA, namely A–U–G. Precisely why the small ribosomal subunit and formylmethionyl transfer RNA bind only to the codon for the N-terminus of a protein chain is not yet clear. When this binding has occurred, a large ribosomal subunit adds on to give a complete ribosome, and synthesis of the protein chain can commence.

The first reaction in synthesis of the protein chain is the linking of the carboxyl group of the N-terminal amino acid to the amino of the second amino acid in the chain. Both these amino acids are bound, by way of their transfer RNA, to adjacent triplets on the messenger RNA, each triplet lying in a groove on the ribosome (Figure 33). The ribosome contains an enzyme *peptidyl transferase*, which catalyzes the transfer of the carboxyl group of the N-terminal amino acid from the 3'-hydroxyl of the terminal ribose of its transfer RNA to the amino group of the adjacent amino acid. The product is a dipeptide linked by its carboxyl group to the transfer RNA of amino acid 2. The ribosome now moves along the messenger RNA in the direction of its 3' end so that the triplet for amino acid 2 now lies in the groove previously occupied by the N-terminal triplet; simultaneously the triplet for amino acid 3 moves into the other groove. In this process the transfer RNA of the N-terminal amino acid is released, and transfer RNA with amino acid 3 attached binds in the other groove. The cycle of changes is now repeated to give a tripeptide linked by its C-terminus to the transfer RNA of amino acid 3. Precise details of how each cycle is brought about are still unclear. A number of enzymes are involved, and energy is obtained by hydrolysis of GTP to GDP.

After a number of these cycles equal to the number of peptide bonds in the protein, the protein chain is complete: it is attached by its C-terminus to the transfer RNA of the C-terminal amino acid, which is itself bound to the C-terminal codon in one of the two grooves of the ribosome. That no more amino acids will be added to the chain is dictated by the adjacent triplet on the messenger RNA: if this triplet is U–A–A, U–A–G or U–G–A chain termination occurs. The protein is hydrolyzed from its C-terminal transfer RNA, and both are released. The chain termination triplets do not bind a transfer RNA but rather a protein, which, in a manner not understood, is involved in release of the completed chain.

The problem remains of how each protein acquires the characteristic conformation or three-dimensional structure on which its biological properties depend (see p. 40). If the one gene–one protein concept is correct, then this conformation must be determined by the same gene that directs the amino acid sequence of the protein. The simplest way in which it can achieve this is if the conformation of the protein is a spontaneous consequence of its amino acid sequence, and evidence suggests that it is. The conformation of a protein is largely stabilized by hydrogen bonds between the carbonyl and imino groups of the peptide chain and by noncovalent bonds between the groups of amino acid side chains. It appears that as amino acids are added to a protein chain at a ribosome they soon become stabilized by these non-

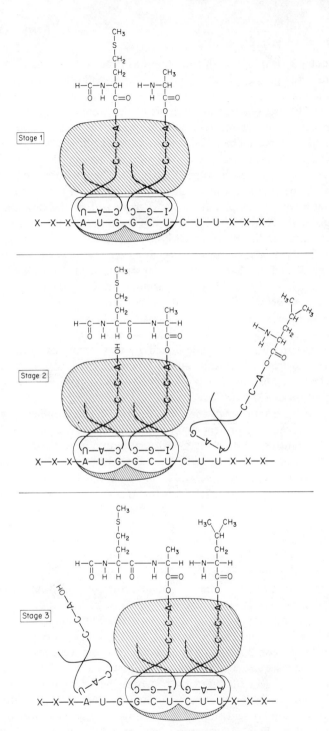

Figure 33. Start of synthesis of a protein chain with the sequence N-formylmet–ala–leu–.

covalent bonds into a specific one of the many conformations that they can assume. (The different *conformations* of a molecule are the different structures it can assume as a result of the rotation of groups relative to one another about single bonds.)

Experimental evidence in support of this belief came first from experiments on pancreatic ribonuclease, the conformation of which is illustrated in Figure 14 (p. 41). This is stabilized not only by noncovalent bonds but also by four covalent disulfide bonds between the side chains of cysteine residues. C. A. Anfinsen and his colleagues at the National Institutes of Health severed the disulfide bonds of ribonuclease by reduction with mercaptoethanol and the protein lost its ability to digest RNA—presumably because the chain partly unfolds, separating groups involved in catalysis. However, after some hours of reoxidation in air the disulfide bonds reformed, and the protein regained its enzymic activity. It is clear that of all the possible ways in which the eight cysteine side chains could have reformed disulfide bonds, they chose the four correct ones. Hence, although the ribonuclease molecules must undergo certain fluctuations in conformation after reduction, noncovalent forces must repeatedly bring molecules back to the correct conformation where the correct disulfide bonds can reform.

F. M. Richards in the same laboratory performed an even more surprising experiment. He found that when ribonuclease was incubated with the bacterial enzyme subtilisin only one peptide bond was severed: that between amino acids 20 and 21 from the N-terminus (see Figure 13, p. 39). The enzyme retained its activity. If the two peptide fragments were separated, neither was active. But when the fragments were merely mixed in solution the enzymic activity returned. It is clear that random collision occurred between the two fragments until they became reunited by the correct noncovalent bonds.

The conformation of a protein can, however, be disrupted beyond return—it is then said to be *denatured*. This shows that the biologically active conformation is not the only stable one. Hence the formation of the correct conformation must depend on its being built up gradually as the chain elongates at the ribosomes. Evidence for this is that some proteins acquire enzymic activity before their release from the ribosome.

Chapter 6

The Regulation of
Gene Activity

I. Jacob and Monod's Theory of Gene Regulation

We have seen that most genes act by directing the formation of a protein. But a glance at higher organisms tells us that this action must be regulated: human red blood cells, for example, make hemoglobin, while other cells do not. In fact the development of a fertilized egg into a differentiated multicellular organism must be founded on the controlled regulation of gene activity. But gene regulation also occurs in single-celled organisms and is valuable for their survival. *E. coli* bacteria, for example, normally only make the enzyme β-galactosidase, which hydrolyzes lactose to glucose and galactose when surrounded by a solution of lactose. Hence in the human gut they do not waste energy making this enzyme unless their host has recently provided them with milk sugar as a nutrient.

It is in fact this particular example of gene regulation that was intensively studied over many years by J. Monod and F. Jacob and their colleagues at the Pasteur Institute in Paris. Their elegant experiments have revealed with great clarity how the activity of the gene in *E. coli* that directs the synthesis of β-galactosidase is regulated, and for this work they were awarded the Nobel Prize in 1965. Induction of β-galactosidase in *E. coli* can be simply demonstrated in the laboratory. For example, 1 liter of a sterile solution of inorganic salts and maltose could be put in a 4-liter flask and a few milliliters of a culture of *E. coli* poured in. The flask is shaken for a few hours at 37°C until optical density measurements show that the bacteria are in the middle of logarithmic growth; that is, their numbers are doubling each time a set number of minutes elapses. The culture is cooled to 0°C to stop growth, and the bacteria are separated by centrifugation. They are then resuspended in 2 liters of the solution of inorganic salts and maltose and again shaken at 37°C. A small volume of lactose solution is then added, and samples are removed every few minutes and cooled; later their β-galactosidase activity is determined. The activity per milliliter of culture is found to increase rapidly from a few minutes after the addition of lactose. If the bacteria are separated and reincubated in a medium with maltose but no lactose, the activity per milliliter remains constant. Certain other compounds of structure related to lactose will induce the formation of β-galactosidase. One is the lactose analog isopropylthiogalactoside, used in experiments because it is effective in smaller concentration than lactose and is not broken down by the enzyme.

The increase in β-galactosidase activity on induction results from synthesis of the enzyme from amino acids: when inducer is present, β-galactosidase comprises a greatly increased proportion of all the proteins synthesized during growth of the cells. In fact, cells formed in the presence of inducer each contain up to 5,000 molecules of β-galactosidase, compared with only about 5 molecules in cells formed without inducer. Two other enzymes are induced with β-galactosidase, and the relative proportions of these three *lac* enzymes always remain constant. The two other enzymes are β-galactoside permease, which is responsible for lactose uptake, and β-galactoside transacetylase, whose function is unknown.

Jacob and Monod's clues as to how the genes that control the synthesis of these enzymes are regulated came largely from their isolation of mutant strains of *E. coli* with defects in this regulation. The following are some of these mutants:

1. Mutants in which one of the three enzymes is not synthesized on induction. Mapping techniques showed that these have defects in three adjacent genes each of which directs the synthesis of one of the enzymes. These *structural* genes are arranged on the chromosome in the order β-galactosidase (denoted *z*), permease (*y*), and transacetylase (*a*).

2. Mutants that make the three enzymes whether or not inducer is present (*constitutive* mutants) and whose defect is always in a region of the chromosome immediately preceding the *z*, *y*, and *a* genes. It therefore appeared that this *i* region was a separate *regulator* gene that confers on the cell the added refinement of only forming enzymes concerned in the metabolism of lactose when lactose is there to be metabolized.

3. Mutants with two chromosomes per cell: one with an active regulator gene but a defective structural gene for β-galactosidase (i.e., i^+z^-) and the other with a defective regulator gene but active structural gene (i.e., i^-z^+). Such mutants only produce β-galacosidase in the presence of an inducer. Hence the intact regulator gene can regulate the intact structural gene on the other chromosome. Hence it must do this by directing the formation of a molecule that can diffuse through the cytoplasm.

4. A clue as to how this diffusible regulating molecule acts is given by mutants of another kind that have the regulator gene intact but nevertheless are constitutive, making the three enzymes whether or not inducer is present. These mutants have defects in a small region of the chromosome (named o) near the beginning of the β-galactosidase structural gene. If another chromosome with an intact regulator gene is introduced into such cells, they remain constitutive. Hence it is clear that mutations at the o position can prevent a perfectly good regulating molecule from acting. This suggests that this molecule acts at the o (operator) position.

From the properties of such mutants, together with some biochemical evidence, Jacob and Monod suggested how the genes that direct the formation of the three enzymes are regulated (Figure 34). They suggested that the

Figure 34. Regulation of genes of the lactose operon in *E. coli*. The β-galactosidase gene contains around 3,500 base pairs, and the other genes are drawn roughly to scale.

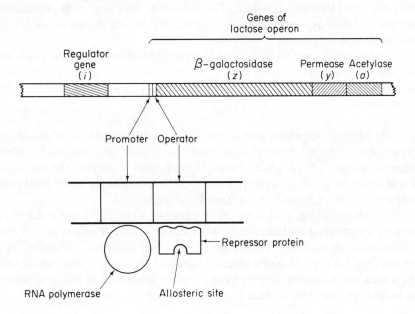

regulator gene directs the formation of a *repressor* protein. This protein has affinity for the sequence of nucleotides of the operator, similar to the affinity of an enzyme for its substrate, and can bind there. When bound to the operator, the repressor prevents the movement of RNA polymerase along the three structural genes: no messenger RNA is formed, and hence the three proteins are not synthesized. However, the repressor protein has another active site with affinity for molecules of inducer such as isopropylthiogalactoside. When an inducer molecule is bound at this site, the conformation of the protein changes, and it loses its affinity for the operator. Thus when inducer is present, synthesis of messenger RNA and hence of the three proteins can proceed.

2. Isolation of the Lactose Repressor

For a few years after Jacob and Monod put forward their theory it was possible to interpret their evidence for the regulation of β-galactosidase in another less elegant way: the repressor acts on messenger RNA to prevent protein formation rather than on DNA to prevent RNA formation. However, in 1966 the mechanism of Jacob and Monod was brilliantly confirmed by W. Gilbert and B. Müller-Hill at Harvard University, who isolated the repressor and demonstrated its binding to isolated DNA.

When isolating a compound from living cells for the first time, it is essential to be able to assay the success of different approaches. Most proteins are enzymes, and the success of their isolation is indicated by the enzyme activity per milligram of protein which will increase to a maximum as purification proceeds. But the repressor is not an enzyme, and some other method of assay was needed. Gilbert and Müller-Hill chose equilibrium dialysis. They decided to put a suspension of broken *E. coli* cells into a dialysis sac and shake it in a solution of the inducer isopropylthiogalactoside that was labeled with radioactive carbon. The small inducer molecules can pass through the pores of the dialysis sac, whereas the large repressor molecules should not. Hence, at equilibrium, there should be more radioactivity within the sac than without, since within the sac will not only be free inducer but also inducer bound to repressor. The excess of radioactivity within the sac will give a measure of the quantity of repressor.

However, the amount of repressor in normal *E. coli* is so small that before purification it is impossible to detect in this way. Gilbert and Müller-Hill therefore isolated a mutant strain of *E. coli* with a repressor that binds inducer more strongly than normal. They selected this from a random mixture of *E. coli* mutants by devising conditions under which only strains that respond to low levels of inducer could grow. Even with broken cells of this strain they could at first detect no binding of labeled inducer within the dialysis sac. But by fractionating the cell proteins by precipitation with different concentrations of ammonium sulfate they at last had success: with one fraction there were 10,400 counts/minute of radioactivity per milliliter of solution within the sac, and only 10,000/ml of solution outside. From this point progress was straightforward. Further fractionation by ammonium sulfate, followed by column chromatography, gave a product that would produce a concentration of radioactivity within the dialysis sac that was twice the concentration outside.

Having devised the procedure on this "tight-binding" *E. coli* mutant, they were able to repeat it on normal *E. coli*. The ability of their product to bind inducer was not destroyed by ribonuclease or deoxyribonuclease but was destroyed by pronase. Hence the repressor appeared to be a protein, and its rate of sedimentation suggested a molecular weight of about 150,000. They verified in a number of ways that it was, in fact, the repressor of β-galactosidase induction. Thus different labeled inducers were bound in proportion to their activity in inducing the enzyme in living cells. Also, they ran through their purification procedure with three mutant strains of *E. coli*: one had the *i* gene deleted and produced no repressor, and the other two produced repressors that bound inducer only very feebly. The first strain gave a product that would not bind inducer, while the other two gave one that bound it feebly.

Having isolated the repressor of β-galactosidase induction in a highly purified form, they were now in a position to test whether it does in fact bind to the operator site on *E. coli* DNA and is released from this binding by inducers, as Jacob and Monod had suggested. To follow the binding to DNA, they grew *E. coli* in the presence of radioactive ^{35}S and isolated labeled repressor. Because there is only one operator site on the whole *E. coli* chromosome, the amount of labeled repressor bound to whole *E. coli* DNA would be too small to detect. They therefore used DNA from a bacterial virus that was known to have the β-galactosidase region of the *E. coli* chromosome attached to it.

They demonstrated the binding of repressor to this DNA by density gradient centrifugation. A glycerol density gradient is a buffered glycerol solution in a centrifuge tube whose glycerol concentration, and hence density, increases regularly toward the base of the tube. When the DNA was placed on the top of such a gradient and centrifuged at 65,000 revolutions/minute for 140 minutes the DNA moved in a band to a position near the bottom of the tube. When the labeled repressor was similarly centrifuged, it remained near the top of the tube. But if a mixture of DNA and repressor was centrifuged, some of the labeled protein was found to move to the bottom of the tube with the DNA. It was proved in a number of ways that this was repressor bound to the operator site. If, for example, the glycerol gradient contained the inducer isopropylthiogalactoside, no radioactivity was associated with the DNA. If the bacterial virus carried DNA from *E. coli* with a defective operator (*operator constitutive mutant*), again no radioactivity was associated with the DNA. The repressor would not bind if the two strands of the DNA double helix were separated by denaturation.

The *lac* repressor has now been completely purified and studied in more detail. It is composed of four identical subunits of molecular weight 40,000. Normal *E. coli* produce only 10 to 20 molecules of repressor per chromosome.

3. How Repressor Proteins Act

It has been seen in the first two sections of this chapter that regulation of the β-galactosidase gene is founded on the binding of a protein to one particular small section of the *E. coli* chromosome. It is now clear that reversible binding by proteins of specific regions of DNA is of prime importance in the regulation of microbial genes. Since the *E. coli* chromosome has about 3.8×10^6 base pairs, the lac repressor probably binds to at least 12 of these base pairs, because the four bases of DNA could be arranged in sequences of 12 in 1.7×10^7 ways and in sequences of 11 in 4.2×10^6 ways. Hence a particular sequence of 11 or less is likely to occur more than once along the *E. coli* chromosome. Twelve base pairs stretch about 35 Å, or more than one complete turn, along the DNA double helix, but the lactose repressor is large enough to bind along this length. Precisely how a protein recognizes a sequence of bases when they lie along the center, rather than the outside, of the double helix is still unclear.

The binding of the lactose repressor to DNA is released when the inducer becomes bound at another site on the repressor molecule. Effects of this type, in which a molecule being bound at one site on a protein causes a change in affinity for another molecule at another site on the protein, are known as *allosteric* effects. They appear to result from the binding of one molecule altering the three-dimensional conformation of the protein and, as a result, its affinity for the other molecule.

How the repressor prevents transcription of the lactose genes is now clearer. RNA polymerase binds to a *promoter* site that immediately precedes the operator site on the *E. coli* chromosome (Figure 34). Mutations at this site, by altering its affinity for RNA polymerase, increase or decrease the rate at which the polymerase forms messenger RNA with the three lactose genes as template. Only when the repressor is released from the DNA as a result of binding inducer can the RNA polymerase move along the DNA. The three structural genes are then transcribed into one length of messenger RNA. When a particular group of genes is always transcribed into a single length of messenger RNA, that group of genes is named an *operon*. Ribosomes become attached to the messenger RNA as it is being formed, and protein synthesis proceeds. Roughly 10 molecules of β-galactosidase protein are synthesized for every 5 of permease for every 2 of acetylase. How this comes about is not yet clear.

Unlike lactose, which is an inducer of gene activity in *E. coli*, certain compounds in the culture medium will repress gene activity. An example is histidine, which represses the activity of the genes that code for the enzymes of histidine synthesis. Hence the organism only synthesizes histidine when it is needed. The mechanism of this repression is remarkably similar to that of induction by lactose. A repressor protein binds to an operator site on the DNA just preceding an operon of the 10 structural genes for the enzymes of histidine synthesis. But the behavior of this repressor is the converse of that of the lac repressor: it only binds to the DNA when histidine is present (rather than absent) at a second binding site.

More facts about the induction and repression of genes in *E. coli* certainly remain to be discovered. For example, the sugar arabinose induces the synthesis of three enzymes involved in its metabolism. The three structural genes for these enzymes are grouped as an operon and preceded by a regulator gene. But when this regulator gene is defective the enzymes are not synthesized either in the presence or absence of arabinose. Hence this regulator gene does not form a repressor similar to the lac repressor. Again, when *E. coli* is incubated in a medium containing glucose and lactose the enzymes of lactose metabolism are not induced. Precisely how glucose acts is still unknown, but mutations at the promoter site of the lactose operon can eliminate this action of glucose, showing that interference with the binding of RNA polymerase to DNA is involved.

4. Gene Regulation by Modified Binding of RNA Polymerase

There is another type of gene regulation that is vital to the correct functioning of bacterial cells. Many genes are not regulated by components of the growth medium but are constantly active; that is, their enzymes are constitutive. But the expression of these genes relative to one another must be correctly balanced. Thus the *E. coli* cell requires hundreds of molecules of each enzyme catalyzing the degradation of glucose but needs only about 20 molecules of the lactose repressor. One factor in this differential synthesis of proteins appears to be that the promoter sites of different genes have different nucleotide sequences and, as a result, different affinities for RNA polymerase. The higher the affinity, the more often will an RNA polymerase molecule that collides with the promoter site be bound and initiate transcription. Clear evidence that nucleotide sequence at the promoter site influences the rate of transcription comes from a study of mutations at the promoter of the gene for the lactose repressor. These mutations can increase the number of repressor molecules, some mutant strains synthesizing around 1,000 molecules per cell. Relative rates of protein synthesis can also be regulated in other ways. One is by the messenger RNAs having different nucleotide sequences near the initiation points of protein synthesis, resulting in different affinities for ribosomes. Another is by the messenger RNAs having different nucleotide sequences at the 5′ terminus (i.e., the end of the molecule with the free 5′-phosphate). This results in different affinities for the ribonuclease that degrades messenger RNA. These two ways of regulating rates of protein synthesis are also founded on the binding of proteins to nucleotide sequences.

There is another mechanism of regulation in bacteria and viruses that allows genes to be expressed in a sequence in time. RNA polymerase is a large molecule of 500,000 molecular weight and is composed of five subunits bound by noncovalent bonds. One subunit is named the *sigma factor* and is essential if RNA polymerase is to bind only to promoter sites: if sigma is absent, RNA polymerase initiates RNA synthesis from many other sites on DNA. Moreover, there are certain promoter sites to which RNA polymerase with sigma will not bind. They can only be recognized by RNA polymerase in which sigma is replaced by another similar polypeptide. The best understood examples of this are in bacterial viruses. For example, when the virus T_4 infects an *E. coli* cell, the *E. coli* RNA polymerase binds to only a few T_4 promoter sites to transcribe the "pre-early" genes. One of these genes codes for a new polypeptide that replaces sigma. The resulting RNA polymerase can bind only to the promoter sites of a second group of "early" genes on the T_4 DNA. Proteins directed by these genes begin to appear around 4 minutes after infection. One of these is yet another sigma-type polypeptide that displaces the previous one to form an RNA polymerase that can only bind to the promoter sites of the "late" genes—genes that are expressed about 10 minutes after infection and that include genes for the phage coat protein and for an enzyme, which lyses the bacterial cell. The time after infection at which the various T_4 genes are expressed is thus founded on two processes. The first is that RNA polymerase takes about 1 minute to form messenger RNA from the average gene so that it is some minutes before proteins coded by genes at the end of an operon are synthesized. The second process is the successive synthesis of sigma-type factors by these genes.

To what extent the successive synthesis of sigma-type factors is made use of by bacteria to control their cell cycle is unclear, but one instance is known. When short of nutrients, *Bacillus* bacteria change into inert spores with resistant coats. Spore formation requires the expression of many genes that are inactive during growth of the cells. RNA polymerase only binds to the promoter sites of these genes when the normal sigma factor is replaced by another polypeptide.

5. Epilogue: The Unsolved Problem of Gene Regulation in the Development of Higher Organisms

Heredity in all organisms is founded on cellular heredity—the ability of cells to form like cells by division. We are, finally, in a position to outline the chemical basis of this process. It will be illustrated by an idealized situation: a bacterial cell that divides to give two daughter cells each of which then grows until it reaches exactly the size and composition of the parent cell. At division, each daughter cell receives exactly half the DNA of the parent and roughly half of the molecules of every other compound. The number of molecules of each compound must now double as each daughter cell grows to the parental size. The DNA molecules double by each forming self-copies by the mechanism of Watson and Crick. Few, if any, other molecules form self-copies. DNA molecules direct the formation of molecules of messenger, transfer, and ribosomal RNA. Messenger RNA directs the formation of protein molecules of identical structure to those already present. Most of these proteins are enzymes, and they catalyze the formation of many other molecules, such as those of carbohydrates and lipids, from molecules of nutrients absorbed from outside the cell. Certain other molecules whose number must also double, such as those of water, are simply absorbed from outside as the cell grows.

We can now also begin to understand how the newly formed molecules become grouped into the correct structures, such as ribosomes and cell membranes. The formation of a hemoglobin molecule from two identical α subunits and two identical β subunits provides a simple illustration. Each of these four subunits is synthesized separately in the cell, and each folds spontaneously into its characteristic conformation, which is stabilized by noncovalent bonds. In one particular orientation two α subunits fit neatly beside two β subunits to give a hemoglobin molecule, again stabilized by noncovalent bonds. The subunits attain this orientation by chance during random collision. There is also experimental evidence that ribosomes are built up spontaneously in the cell in a similar way; the small subunits of *E. coli* ribosomes can be disintegrated into RNA and 19 component proteins that, under different conditions, reassemble into normal subunits.

The inheritance of Mendelian unit characters by multicellular organisms is founded on cellular heredity of the kind we have outlined for bacteria, but modified by differentiation. In these organisms cells usually divide to give daughter cells that differ slightly from one another in chemical composition. Nevertheless the inheritance of Mendelian unit characters can still be understood in terms of cell molecules. Red and white flower color in peas, for example, results from a restricted group of cells in the adult organism forming, or not forming, a colored pigment. Each of these cells contains a pair of genes that determine whether the flowers will be red or white. These genes must have been replicated at every cell division between the fertilized egg and the cells of the flower but have only become active in the mature plant.

A query arises in this connection. We have concluded that the function of most genes is to direct the formation of a specific protein. Are we to assume that the form in which every Mendelian unit character appears is the result of the formation of a protein of specific structure? Does a pea plant produce flowers at the end of the stem merely because a protein of a certain chain length and sequence is formed in certain cells, and produce flowers along the stem when this protein is absent or its sequence is changed? The answer is almost certainly yes, for it is difficult to conceive of DNA having more than the two primary functions that have been discovered, namely, to direct its own self-copying and to direct the formation of RNA, which, if it is messenger RNA, will in turn direct the formation of proteins. Dominant genes, therefore, are those that direct the formation in certain cells of an organism of protein molecules of a certain chain length and sequence. Recessive genes are those that direct the formation of no protein or of a protein of markedly different sequence. When both genes of a pair are recessive, so that the protein that produces the dominant form of the character is absent, the recessive form appears. In a few instances there is direct evidence that these conclusions are correct. For example, Mendel's dwarf peas differ from normal peas in lacking the growth factor gibberellic acid and hence probably lack an enzyme involved in its synthesis.

The task of this book—to demonstrate how inherited differences between one living organism and another are founded on differences in the structure of chemical molecules—is therefore completed. But one large gap in our understanding of chemical genetics remains: the development of a fertilized egg into a differentiated multicellular adult is founded on the controlled regulation of genes, but the mechanism remains obscure.

Changes in gene activity during development must be induced by changes in composition of the cytoplasm surrounding the chromosomes, and differences in gene activity between the cells of a developing embryo must first arise from different sets of chromosomes entering different regions of the egg cytoplasm. Later in development, differences in cytoplasmic composition among the cells of an embryo, which in turn induce changes in gene activity, can be produced by differences in cell environment.

How do components of the cytoplasm activate certain genes and repress others, and what is the chemical structure of these components? Their initial action is almost certainly to bind at the allosteric site of a protein that is more directly responsible for the change in gene activity. It was once expected that they would have unique chemical structures with the sole function of inducing particular genetic changes, but search for such embryonic inducers has been unsuccessful. On the contrary, embryonic development is found to be induced by changes in concentration of components that are by no means unique, such as oxygen and sodium chloride. This need not, however, surprise us, since familiar metabolites regulate allosterically the activities of the enzymes of intermediary metabolism in mammalian cells.

If embryonic inducers do, in fact, regulate allosteric proteins how do these proteins in turn regulate gene expression? It seems most likely that they control the transcription of certain genes into messenger RNA. Some may do this like the repressor protein that controls the transcription of the lactose genes in *E. coli*, as is suggested by studies of hormone action. The rate of synthesis of certain enzymes by animal cells increases when certain hormones are present and decreases again when they are removed. The simplest explanation is that each hormone binds reversibly to an allosteric protein whose affinity for an operator site on the genetic DNA is thereby changed.

However, certain features of gene regulation in higher organisms are so different from those in bacteria that different mechanisms must be involved. For example, during development certain genes on a chromosome can be activated in a way that is self-sustaining: the same genes tend to remain active on all chromosomes formed from the original one despite changes in environment. Thus many adult tissues can be disintegrated into their component cells, which can be grown in culture. These cultured cells and their progeny continue to make proteins characteristic of the tissue: cartilage cells make collagen; pituitary cells make hormones. Hence their pattern of gene expression does not depend on a continuous supply of inducing agents from the correct surrounding tissues but is sustained when the cells are merely bathed in the simple components of the culture medium.

Not only gene activities but the ability of genes to become active in a certain environment can be sustained over many cell divisions in foreign surroundings. This is most arrestingly demonstrated by recent experiments of E. Hadorn at Zürich. Larvae of *Drosophila* have different regions or *disks* of undifferentiated cells that each give rise to different organs in the adult, such as sex organs, legs, head. These disks can be removed and implanted into the body cavity of other *Drosophila* larvae or of adult *Drosophila*. When the larvae become pupae, the implants differentiate into the correct adult structures: a disk in the body cavity can, for example, form an eye. But in adults, although the cells of the implant divide, they remain undifferentiated. Adult flies only live for a few weeks, but part of the implant can be transplanted to another fly and then to another. Hadorn has in this way kept the cells of disk tissue growing and dividing for many years. If part of this tissue is implanted in the abdomen of a larva, it later differentiates into adult tissue and normally into the adult tissue for which the original disk was determined (though occasionally, and interestingly, into different tissue).

Genes on a chromosome can also be inactivated in a way that is self-sustaining. This is most clearly demonstrated in female mammals. Early in development, one of their two X chromosomes is inactivated in cells that will form body tissues. All chromosomes derived thereafter from this chromosome are also inactive. This can be proved by studies of X-linked genes. For example, women can be found who have alleles for glucose 6-phosphate dehydrogenase of slightly different amino acid sequence on their two X chromosomes. Each of their skin cells produces only one of the two enzymes, and neighboring cells formed from the same embryonic precursor cell produce the same one.

What mechanisms of gene regulation can account for these peculiar features? The possibility that different patterns of messenger RNA synthesis in different tissues might result from the differential destruction of genes is excluded. For example, John Gurdon and his colleagues at Oxford have injected nuclei from the skin of adult frogs into enucleated frogs' eggs, which have then proceeded to develop into normal tadpoles. It is clear that although many genes are inactivated in adult tissues they are not destroyed.

Another possibility is that genes might be selectively and irrevocably amplified during development of each tissue. Possibly two active copies of a gene might produce no significant effect on a cell, and synthesis of significant quantities of the corresponding protein might require thousands of copies. An instance of gene amplification is known: the DNA of the genes coding for ribosomal RNA is selectively copied around 1,000 times in the frog oocyte, the copies being released within the nucleus. Moreover, the genes for ribosomal RNA are always repeated some hundred times in the chromosomes of every cell of the frog, and the identity and number of these repeats appears to be maintained by some mechanism not found in bacteria. However, amplification does not appear to be essential for significant gene expression: experiments suggest that cells that are actively synthesizing silk proteins or hemoglobin contain only two (or a few more) of the corresponding genes.

A further possibility is that the structure of the DNA of genes might be selectively modified in a way that would set RNA synthesis in a certain pattern. For instance, groups of inactive genes might lack a nucleotide sequence in their DNA to which RNA polymerase binds. At a certain stage of development a protein might be synthesized that would introduce this sequence and, thereafter, these genes would always be replicated with this active structure. Some experiments do in fact hint that the DNA coding for proteins peculiar to differentiated cells differs from that coding for proteins common to all cells. When bromodeoxyuridine is added to the medium of differentiated cells in culture, bromouracil is incorporated into newly synthesized DNA in place of thymine. The cells continue to grow and divide (and hence retain enzymes common to all cells) but often lose proteins peculiar to their differentiated state. When bromodeoxyuridine is removed, synthesis of these proteins is resumed.

It is clear that the mechanism of gene regulation in higher organisms is a mystery whose solution is an important task for molecular biologists over the coming years.

Further Reading

J. M. BARRY AND E. M. BARRY, *An Introduction to the Structure of Biological Molecules*. Prentice-Hall Inc., Englewood Cliffs, N.J., 1969. A discussion of protein and nucleic acid structure is included.

W. HAYES, *The Genetics of Bacteria and Their Viruses*, 2nd ed. John Wiley & Sons, Inc., New York, 1969. The most authoritative and comprehensive account of the subject.

H. ILTIS, *Life of Mendel*. Trans. E. and C. Paul. London: George Allen and Unwin Ltd., 1932. A fascinating account of Mendel's life by a scientist from Brünn who spoke to many of his relatives and friends.

E. SCHRÖDINGER, *What Is Life?* Cambridge University Press, New York, 1945. A consideration of the living cell by a distinguished physicist; it influenced the development of molecular biology.

N. T. SPRATT, *Introduction to Cell Differentiation*. Van Nostrand Reinhold Company, New York, 1964. A useful introduction to animal development.

M. W. STRICKBERGER, *Genetics*. The Macmillan Company, New York, 1968. An excellent comprehensive textbook.

A. H. STURTEVANT, *A History of Genetics*. Harper & Row, Publishers, New York, 1965. An authoritative account by a distinguished geneticist.

A. H. STURTEVANT AND G. W. BEADLE, *An Introduction to Genetics*. Dover Publications, Inc., New York, 1962. A reprint of a clear account of classic genetics written by distinguished contemporary workers.

J. D. WATSON, *The Double Helix: Being a Personal Account of the Discovery of the Structure of DNA*. Atheneum Publishers, New York, 1968. An amusing and absorbing account.

J. D. WATSON, *Molecular Biology of the Gene*, 2nd ed. W. A. Benjamin Inc., Reading, Mass., 1970. A beautifully clear comprehensive account of chemical genetics.

H. L. K. WHITEHOUSE, *Towards an Understanding of the Mechanism of Heredity*, 2nd ed. Edward Arnold & Co., London, 1969. A textbook of genetics with its historical development particularly well analyzed.

Glossary

ALLELE. Alternative forms of a Mendelian character *or* alternative structures of the gene controlling a character.

ALLOSTERIC EFFECT. When an atom or molecule being bound at one site on a protein causes a change in affinity for another atom or molecule at another site on the protein.

ANTICODON. Three adjacent nucleotides in transfer RNA that bind to a codon of three adjacent nucleotides in messenger RNA during protein synthesis.

BACTERIOPHAGE. Bacterial virus.

BASE PAIRING. Hydrogen bonding between the bases of two nucleotides. When two nucleic acid molecules bind together to form a regular double helix, adenine is usually paired with thymine or uracil (and vice versa) and guanine with cytosine (and vice versa).

CHIASMA (Greek for "cross"). A cross-shaped junction formed during meiosis between chromatids of paired homologous chromosomes. It is generally agreed to result from breakage of each chromatid and rejoining of the parts of one with the parts of the other. (See *Crossing over.*)

CHROMATID. Before cell division each chromosome is visible as two parallel daughter chromosomes. Until these separate from one another, they are called chromatids.

CHROMOSOME. Proposed in 1888 for threads that appear in nuclei before cell division.

CISTRON. An attempt made in 1957 to introduce a term for functional units of heredity founded on an unambiguous practical test: two recessive mutations are in the same cistron if there is no complementation (reversion to the unmutated characters) when they are introduced into the same organism on separate chromosomes. Although the test turned out to be ambiguous, the term came to refer to a segment of the genetic material that directed the formation of one polypeptide chain. *Gene* is now used instead to refer to this.

CODON. A sequence of three nucleotides in one of the strands of a DNA double helix, or in a molecule of messenger RNA, that determines which of 20 amino acids will be inserted at a particular position in a polypeptide chain.

COMPLEMENTARY NUCLEOTIDE SEQUENCES. Two nucleotide sequences in RNA or DNA molecules that are related to one another by the base-pairing restrictions of the double helix.

COMPLEMENTATION. A classic test to discover whether two diploid organisms with recessive mutations in the same character have mutations in the same gene, i.e., to see whether the mutations are allelic. Individuals are bred that have one copy of each mutant gene on different chromosomes. Only if the mutations are in different genes (i.e., nonallelic) will the individual show the character in its normal form since each mutant gene is then "complemented" by a normal dominant gene on the homologous chromosome. Complementation tests can be made in haploid microorganisms by introducing two chromosomes into one cell.

CONFIGURATION. Best restricted to refer to the fixed arrangement in space of the groups of a molecule, e.g., about an asymmetric carbon. However, sometimes describes what is defined as *Conformation*.

CONFORMATION. The different conformations of a molecule are the different structures it can assume as a result of the rotation of its component groups relative to one another about single bonds. Protein and nucleic acid molecules in the cell are mostly stabilized by noncovalent bonds into a particular conformation on which the biological activity depends.

CROSSING OVER. Introduced in 1912 to refer to recombinations of linked characters. Also used to refer to exchanges of segments between homologous chromosomes that give rise to recombinations.

DEGENERATE CODONS. Two or more codons that specify the same amino acid.

DELETION. Loss of segment from a chromosome.

DIPLOID (from the Greek *diploos*, "double"). Cells with two homologous sets of chromosomes.

DOMINANT. Introduced by Mendel in 1866. A species can have different strains or races that always exhibit a unit character in different forms, e.g., Mendel's pure breeding red- and white-flowered peas. When two individuals from two of these strains are crossed, their progeny usually all exhibit one of the two forms of the character—the *dominant* form. The other *recessive* form can reappear in progeny of subsequent generations.

DOUBLE HELIX. The helical conformation of two DNA or RNA molecules united by hydrogen bonds discovered by Watson and Crick.

GENE. Introduced in 1909 for the, then hypothetical, units of heredity each of which determines the development of a particular character. Now that the mechanism of heredity is more clearly understood, the term usually refers, more precisely, to the segment of the genetic material that directs the formation of one polypeptide chain.

GENETICS. Introduced in 1907 for the science of heredity.

GENOME. The total complement of genes of an organism.

HAPLOID (from the Greek *haploos*, "single"). Cells with a single set of chromosomes. Originally referred to cells of higher organisms but is now also used to describe microbial cells. A bacterial cell sometimes contains several chromosomes but is still considered haploid since these are derived from a single parent.

HELIX. A corkscrew shape.

HETEROKARYON. Cell containing more than one nucleus derived from different individuals.

HETEROZYGOUS. A diploid cell is heterozygous for a particular gene when it contains two different alleles—rather than two copies of the same allele, when it is *homozygous*.

HOMOLOGOUS CHROMOSOMES. Chromosomes similar in appearance that pair at meiosis before separating into different germ cells.

HOMOZYGOUS. See *Heterozygous*.

In vitro (Latin, "in glass"). Describes experiments in laboratory apparatus on parts of disintegrated living organisms.

In vivo (Latin, "in life"). Describes experiments on intact living organisms.

LINKAGE. The tendency of alleles for different characters to remain associated from one generation to the next.

MEIOSIS. Two successive nuclear divisions by which haploid germ cells are formed from diploid cells.

MESSENGER RNA. RNA, transcribed from a gene, that directs amino acids into the correct polypeptide sequence.

MICROSOME. Fragments of endoplasmic reticulum with attached ribosomes isolated from disintegrated cells of higher organisms by high-speed centrifugation.

MUTATION. Originally referred to an abrupt change in form of a character that could be inherited by the progeny of the individual in which it occurred. Now also refers to an inherited structural change in a gene that results in a change in a character.

NONSENSE MUTATION. Mutation that changes a codon specifying an amino acid into one not specifying any amino acid.

OPERATOR. A site on a chromosome where a repressor protein can bind and prevent transcription of adjacent genes by RNA polymerase.

OPERON. A group of genes that is transcribed into a single length of messenger RNA.

PEPTIDE. A peptide bond is a covalent bond between two amino acids formed by elimination of water between the α-carboxyl group of one and the α-amino group of another. Can also refer to the products of this reaction: dipeptides formed from two amino acids linked in this way, tripeptides from three, and polypeptides from several or many are all collectively known as peptides.

POLAR MUTATION. Mutation in a gene that reduces the expression of genes in the same operon that are transcribed (and translated) later.

POLYPEPTIDE. A chain of several or many amino acids linked by peptide bonds.

POLYSOME (or polyribosome). Structure formed when several ribosomes are bound to messenger RNA during protein synthesis.

PROMOTER. A site on a chromosome where RNA polymerase binds and initiates RNA synthesis.

RECESSIVE. See *Dominant*.

RECOMBINATION. The appearance in an individual of alleles for different characters that were not present together in either parent.

SUPRESSOR MUTATION. Mutation at one site on the DNA of an organism that wholly or partly cures (i.e., suppresses) the effect of a previous mutation at a different site on the DNA.

TRANSCRIPTION. The transfer of a particular nucleotide sequence in DNA to a complementary sequence in RNA by means of RNA polymerase.

TRANSDUCTION. The transfer of genes from one bacterium to another by a virus.

TRANSFORMATION. Change in genetic character of a cell produced by its absorption of chromosome fragments released by another cell.

TRANSLATION. The formation of a particular amino sequence in a polypeptide by use of the nucleotide sequence of messenger RNA as a code.

WILD-TYPE. The normal form of a genetic character.

Index

135